THE APP

A TIME TRAVEL APPLICATION

STEVE BAZSULY

This work of non-fiction has names, places and incidents that are supported by documents, emails, news releases and the author's firsthand experiences.

 The address of TBT Enterprises is 5532 Courtyard Drive, Pompano Beach, FL 33063

Dedicated to Sue and our wonderful family.

CONTENTS

CHAPTER ONE

Johnny felt his heart rate increase, the cold sting of death rushing over his skin and raising the hairs on it. As he stared at the TV, the background sounds drowned out, and the only thing he could hear was the voice of the TV reporter.

"At around twelve forty five P.M., it was confirmed that seventeen year old female, Piper Linn vanished from her home, and has yet to be seen since then. Police are ramping up the search for seventeen year old Piper, who is now missing for one week. According to eye witness, her mother, Laura Linn; said she never left home, and was in her room all through the time. Concern for Piper is continuing to grow as it has now been a full week since she was last seen. Detective Inspector Garrett Cornwall said "Enquiries are continuing in our efforts to locate Piper. And that there is possibility that her disappearance could be linked with the nine other cases of vanished people that happened within the month. The signs left behind are indicating so; which includes, a flashing mobile phone, and power outage in the vicinity of the crime scene. He is asking members of the public to come forward with any clues they might have. He also stated that, the police departments were supporting friends and family of the missing people, and is appealing to the missing people to contact home so that they can know that they are safe and well. Friends and family face an anxious wait for her return, or to find out what has happened to her. We are currently connecting to Reporter Leona

Penwortham who is currently reporting life from the community of the last missing person.

"Thank you Dahlia. Piper was last seen in her room around 7pm last Monday, and officers believe she may have travelled to Wile, South Hale. The young lady is white, and of slim to medium build, approximately five feet three INS tall, and has light brown hair which reaches the middle of her back. She is thought to be wearing a dark hoodie, and blue jeans. Police are calling for her to get in touch with her worried family to confirm she is safe, and help her return home. Detective Garrett said that, the time of year is especially important for families, and they are keen to make sure that she is okay. They also would ask that anyone who is with her should encourage piper to contact her family.

"Piper if you are watching this, we want you to know you are not in any trouble. Please contact us, or someone you trust, to let us know you are safe." The Reporter added.

"What are you doing?" A voice from behind broke through the fog surrounding Johnny, causing him to spin around so fast, he almost had a whiplash. His face was already ashen; and sweat had matted the hair around his temples. "What…" He said almost inaudible, disconcerted and confused.

"I asked what you were doing." Billy Martin asked; who was a longtime friend, and partner of the teenager that stood dazed. "What's wrong?" He asked again, noticing the pale palette of his best friend. "Are you alright?"

“Another one disappeared.” Johnny responds, his eyes darting from side to side, before finally he looks up and stares directly at his friend. “Another person vanished.”

Billy felt fear, and for the first time since he entered the room he could wholly understand why his friend was unsettled. “It doesn’t mean anything… it could be a common disappearance case… it could be… you know… just a coincidence. This is freaking bullshit.” He rambled on, running his hand through his hair as he paced few inches away.

“I hope so too, but it is the tenth person. The tenth vanishing case since the time we launched…” Johnny trailed off, unable to complete his words. “It is tenth vanishing case since we launched the app.” He completed, struggling to speak pass the large lump that had settled in his throat.

“I say that’s bullshit!” Billy snapped, stopping to face Johnny who was just as distraught.

“The report says the same… the signs are the same … a flashing mobile phone like a virus attack, outage of power in the vicinity, and… a missing person. It is too much of a coincidence.” Johnny said, saying the words Bill was not ready to hear.

Almost as if the sun of acceptance dawned on them, Bill slouched, moving to the empty sofa that was opposite the place he stood. As he plopped down on it, he felt extremely weak and tired at that moment. It was like the box of Pandora was finally open; a dam had burst out, and there was no way to stop it at that time. “What should we do?” He asked in a tiny voice, rubbing his right hand down his face

“We need to go back. Go back to the time everything started, and stop the whole process before another person gets stuck in a void created by the loop in time.”

Both young men felt emptiness in their heart, and it was no surprise why that was happening. They were planning on shutting down something they had built their whole lives, their dreams, their blood, their sweat and tears. A project that had materialized from their childhood imaginings, a time when they were only boys filled with youthful innocence, naïve dreams, and unbelievable aspirations – but it had come true. And now, they were throwing it all away. All they felt was emptiness; an emptiness that nothing else could fill.

CHAPTER TWO

FOUR YEARS AGO

It was summer; times when school was out, and kids were the happiest. Johnny Jones and Billy Martins were no different. The twelve year olds had been looking forward to this break for a long time. "Mom, we are back." Johnny yelled at the doorway, dropping his backpack, and racing up the stairs.

"Good afternoon Mrs. Jones." Billy yelled after, also dropping his bag and racing after his best friend who was running up the stairs. The excitement mixed in with heat rolled off their bodies as they raced towards an unknown destination. At the very end of the corridor, Johnny flung open the door to the last room. The insides revealed a typical bedroom design of a young boy; painted in blue, and had the additional elements one would expect.

"Come on Bill." Johnny yelled, beckoning to his partner in crime who was also excited.

"I am here, alright." He grumbled, closing the door behind him. "Where do we start from?" He asked, falling back with abandon on the bed.

"I don't know." Johnny replied, also bouncing back.

"We only have three more days before summer break is over." Bill groaned, putting both of his hands behind his head. "Oh man…"

"Ugh don't remind me. And then we get school." Johnny added, saying the word *'school'* in the most disgusting tone he could think of. At his voice, they both burst into laughter, their youthful giggles echoing, and bouncing across the room.

Slowly, their laughter died out, and all that was left behind was silence.

"I wish summer repeats itself…"

"Yeah…" John yawned, thinking about all the fun things that had done during their break. "If only we can go back in time… like a time machine or something, like a cool machine that makes summer happen again." He said enthusiastically, turning to look at his friend.

"Yes, and then we don't ever have to go back to school, but forever be in summer!" Bill agreed, jumping up to his feet, and causing the mattress to bounce. "What if we built the machine by ourselves and become inventors and go back to whatever time we want?"

"Coool… like that time at Anna's birthday party, and all the cool things we get to eat again?"

"Yah, and then knowing the answers to Mr. Rice boring test questions, and getting A on everything. We are like cool villains"

"Let's build a time machine." Johnny repeated, also jumping to his feet. Excitedly, they both jumped on the bed, sharing ideas on what cool features they would add, and what they would do when they go back.

"Johnny Mathias Jones, you better not be jumping on your bed again." Mrs. Jones warning voice came from downstairs. Guilty as charge, both boys got down, and giggled at being caught. Neither of them knew that something that had started out so small and unbelievable would turn out to something feasible and true. Neither of them knew what was coming.

"Uhm… John, what do we need to build a time machine?" Bill asked, a frown burrowing between his brows.

"Wood and roofing sheets… I guess… right?" John answered, scratching his head and mighty unsure of how to start.

"Let's ask my dad when he gets back." Bill said, his stomach rumbling almost immediately. "I am hungry." He giggled.

"Same… let's go back downstairs."

Their excitement was temporarily dismissed, but that was the beginning of a long line of projects yet to come; THE very beginning of THE APP.

CHAPTER THREE

Back to the present

Outside of the police station, the sun stood gloriously above, shining brightly with its golden rays tinting the light blue sky that spanned as far as the eye could see; a very good day if one could say. Though the weather was great, it did nothing to reflect the seriousness that went on inside the station. In one of the rooms, the mood was dark, and the messy appearance of strewn papers, case files, and the LED lighting did nothing to improve the mood, or the tension that filled the air-space. Detective Key stared at the four pictures sprawled out in front of him, his eyes darting back and forth between them. No matter how much he stared, nothing seemed to make sense at all. Nothing at all.

"Still at it?" The familiar voice of his partner drew his attention, causing him to turn around. "Yep… still at it." He sighed.

"Here, this should help." Detective Maxwell said, handing over a coffee cup.

"Thanks." Key thanked, sipping from the steaming Styrofoam cup. As soon as the caffeine hit his blood stream, he couldn't hold back a contented sigh that managed to escape through his mouth. There was no way he could survive life without coffee. "Back from rounds?"

"Yes… I also learnt nothing. No one saw her that day. According to her mother, after the outage, she went up to her daughter's room to check if she was okay, but she was not there. She first thought she had left her room to look for her, but no matter where she searched, her daughter was nowhere to be found." Maxwell responded, staring down at the pictures on the table. "It is like every one of them vanished."

"I know… which is why the whole case seems too strange. According to all of them, they never left home… and yet – they are nowhere to be found."

"And the clues… none of it makes sense, it is not like it is an alien invasion or something." Maxwell sighed. "Is there anything from the Technical analogist?"

"No… nothing yet. He said he would call by three p.m. that's like fifteen minutes from now. Let's hope he learns something new." Key replied, running a hand through his hair. He had worked on so many cases, some brutal, some complicated… but this – was so strange, too strange and clean. He was sure of one thing though – whoever it was, that was indirectly connected to this case was an expert; an expert in kidnapping… or a serial killer. The buzzing sound of his phone broke into his thoughts, and on the very screen, the name of the Technical analogist was displayed. "He is calling earlier than I thought." He said, catching the attention of his partner. Putting the phone on loud speaker, he dropped it on the table. "What do we have?" He asked.

"I have good news, and bad news… what do you want to hear?" The voice on the phone asked.

Key looked at his partner's face, and then back at his phone. "Good news." He replied.

"The good news is I was able to connect three major interfaces from the phones retrieved from the crime scenes. Reaching DNS, and using Arduino IDE, Linux-based tool and mirroring pHreaker code hack, I was able to access the VPN radio network, using digital modulation. Three modulations had its IP address hidden from all incoming traffic and any networks..."He explained. "But that's not all, AI advanced global sniffing algorithm code was mixed with the retrieved algorithms '*Compaq UFI*, to walk through or you could say… map out the pattern directory."

"So did you get it?" Key asked.

"Yes, we were able to create a new task and set the current options x. while paying special attention to the data option y. This is where we replace the vulnerable value in the post data with an asterisk. The asterisk is a special notation in wall-map that says we still need to set one more option before we can start the task. We need to set the correct content type and SOAP action in the HTTP headers in the request. Otherwise, the server will just return with 500 errors. This is what the next part of the method does, as detailed in listing 9-33."

"English please" Maxwell cut in, causing a groan from the one on the other side of the phone.

"Sorry about that. What I meant was, there is software I used which works as a reverse-engineered ethical hacking code to breach through, and pick out similar… Uhm… similar things between all the phones we brought into the lab. Each and every one of them had an almost identical LEAR, except their IP address was different, but it is definitely connected."

Key threw a small celebratory fist in the air, knowing this was the first real lead they had since the whole investigation began. "That is very good, very good!" He repeated.

"About that..." The voice cut in, hesitation apparent. "The bad news is, even though we were able to connect all the interfaces together, the last line of URL leads to a blank page... like nothing. Almost like someone concocted all these massive lines of codes for nothing, which is very strange.

"Is this a cyber crime....?" Max asked.

"It might be, it might not be..." The voice responded. "It doesn't seem like an attack. It just seem like it exists, but at the same time, it doesn't."

"Alright, thank you. I will get back to you later." Key said, cutting the call as they said their goodbyes. A new layer of frustration was added to his chest, and he could feel the beginnings of a new headache. They gained something, but were left with nothing; what else was more frustrating than that?"

"What do we do now?" Maxwell asked, sipping out of his own cup of coffee.

"I wish I knew the answer to those too... how ten teenage children vanished within a month...This is probably the hardest case I have gotten on. The public, their family, their friends... everyone wants answers... but we have none."

CHAPTER FOUR

Back to the past

Two years ago.

Fourteen year old Bill leaned back against the chair, his head pounding. He needed food, a good shower, and sleep. He had not gotten any in the past twenty four hours; but he was not the only one facing the same situation. His best friend, Johnny, was also stuck in the same situation as he was..

"Nothing seems to add up no matter how much we change the algorithm. This was stupid all along. There was no way it could ever be possible." Johnny grumbled, kicking the leg of the table. As soon as the shock reverberated off the table's leg, the table shifted back, causing the computer to move to the side.

"What are you doing? Stupid! Are you trying to destroy the computer?" Bill snapped,
"I am not stupid, you are stupid." Johnny retorted. Teenage angst, frustrating situation; no food, no sleep, the fireworks were ready to explode. "You know what? I am leaving. This is…"

"Go now…" Bill said, cutting short Johnny's words by pushing him by the shoulder.

Johnny pretended he was about to leave only to turn around at the very last minute, throwing a punch at Bill's nose.

"Fuck!" Bill cussed, bending over as he held on to his nose that was now bleeding. Angry, and looking for a way to lash out, Bill threw a punch at Johnny, hitting him square in the jaw; but Johnny wasn't about to let the chance go. Doing a fake, he turned the other way and punched hard at the Bill's stomach. *'Mis-direction...'* he muttered.

"Ugh..." Bill grunted out in pain as Johnny's fist connected with his gut, knocking the air out of his lungs. It was a fight for who came out on top, and Bill didn't want to go down alone. Pushing against his midribs, Johnny felt his knees buckle against his will, but he managed to block the next hit with his right forearm even though he was about to keel over. Ducking in a split second, he raked over Bill with his leg, causing both of them to fall heavily to the floor; which in turn knocked Johnny over. Bill's weight pinned Johnny to the ground, temporarily immobilizing him.

Reaching a stalemate with no way out, both boys held on to each other's cloth, their prides not letting the other go. "Let go." John said through gritted teeth.

"No, you let go." Bill replied, tightening his grip on John's cloth.

Huffing and puffing, almost anonymously, they both let go in a split second, rolling over and staying inches away from each other.

"Thank God my phone was not in my pocket, you would have broken it" Bill mumbled. At his words, John turned towards him. "What?" Bill asked, rubbing his bruised chin.

“What did you just say?” John asked.

“I said, if my phone was with me, you would have broken it, stupid!”

“That’s it!” John burst out excitedly, startling Bill.

“What is wrong with you?” Bill snapped as he stared weirdly at his best friend. He wondered if he had hit his head earlier.

“The phone app! That was it! That was the answer we were looking for all along!” John said excitedly, pushing himself up.

“What are you talking ab…” Bill trailed off as he realized what John was all about. “The phone app.” He said incredulously, also standing up. “We can try making the portal like an app.” He laughed, but almost immediately winces out in pain. “You punched me so hard.” He complained.

“You punched me so hard too.” John protested, rubbing his rib area. “Where did you learn how to fight like that?”

“I can ask you the same…” Bill said. Almost immediately, they burst out laughing. “We need to get back to work.” Bill said, shaking his head.

CHAPTER FIVE

Back to the past

One year ago.

"Mobile applications have become highly pervasive in recent years. So, the quality of the app we are producing is essential since application failures can lead to serious consequences, such as algorithm loop; and then we would have to start all over again." Fifteen year old Johnny stated. John looked up as Bill began to chuckle, causing him to raise a brow. "What's funny?"

"Once upon a time, you thought building the key to the past will happen with wood and roofing sheets." He smirked.

"Like you were any different." John scoffed, turning back towards the computer screen. They had come a long way in their research. Three good years had passed since they embarked on this project, and slowly and steadily, they began to see results. "Will you please get back to work? We still need to complete the essay assignment."

"Is that due tomorrow?" Bill groaned.

"Yep… so, be fast."

In silence, both teenagers continued to tap away on the keyboards, different codes appearing on the screen.

"I created clustered views to answer the research questions and used existing surveys to complement the algorithm that stops the time for a few minutes." Bill said, his face illuminated by the computer screen in front of him, his eyes blinking as the loading code moved from ninety eight to a hundred. "Finally…" he whispered, pressing in some more keys.

"Remember that, in most cases, mobile applications are developed to be an interface to the standard application. The mobile application should sit between the standard application and the mobile client, and it should handle communications between the mobile client and the standard application; which means the past and the present." John explained.

"But don't you think a loop would appear, and some parts would be skipped out?"

"There are, of course, exceptions where a mobile application is developed independently, but the security controls will remain the same."

At the doorway, Mrs. Jones rolled her eyes at the two young boys that were stuck at their computers, discussing things she couldn't understand at all. She had always wondered why they preferred to stay indoors all day; not doing normal things boys their age did, but she had given up when she realized they were only, truly happy when they were tapping away at the keyboards, doing whatever they always did. And besides, their grades didn't drop, but improved with time.

Clearing her throat to signify her presence, the two teens turned towards the doorway, shock written on their face.

"Mom, what are you doing?" John grumbled, rushing to his feet, so he could close the door and hide what they were doing. "I told you to always knock." He said.

"The door was wide open." She replied.

Scowling, John turned to where Bill was seated. "You didn't close the door?" He asked.

"I forgot." Bill replied, a sheepish look on his face.

"Either ways, here." She said, handing over the tray that held a plate of cookies and two cups of milk.

"We are fin-"

"I am hungry." Bill said, cutting short John's words, and earning him another scowl.

Pleased with herself, Mrs. Jones stretched her hand out, placing the tray gently on her son's waiting hands. "Remember to wash the dishes." She said, then she was gone.

Three hours passed, and the boys were still hard at work. With the cookies and milk gone, extra codes written, and more algorithms formed, Bill was getting restless. He dialed into the PTP's code, and began correcting the information on it. Concentration at the peak with no one saying anything, it didn't take long for him to crack half of the

PTP's code "Who, who, whooo!" he sang excitedly, his arms up in the air as he danced to the off tune.

"I think I found a way to extend the time one can go back into the past."

John paused what he was doing, staring bewilderedly at his best friend. "You are lying."

"No, I am not. Look at this." He said, turning the screen towards him. As soon as he did that, the code changes its dynamic, skipping a part, and looping out an extended version. "No, no, no…" He cried, turning the screen back to himself, as he tapped some more info on the keys. "I swear, I had a clear path just a moment ago. And now a bug has appeared." Looking up, he found Johnny with a raised brow that showed he wasn't impressed. "I swear, I am not lying. I got excited too early."He groaned. He tapped on a few keys but the curved path wasn't budging, and he knew the puzzle had to be cracked, so he could get into the other code lines. "How am I supposed to get this…" He sighed, turning to look at John once more.

"Fine, let me see what I can do. Let's change seats." John said. As he began to tap away, he realized that there was a probable code beneath it. "You are right, there might be a way to extend the time once more." He reluctantly agreed.

"You see." Bill said smugly. "I told you I got something earlier."

"One of the biggest mistakes that mobile application developers make is assuming that only mobile devices will interact with the mobile application. Assuming the mobile application server is network accessible, any system with access to the network will be able to attack that application server. In order to provide this level of access, the mobile application server will be Internet accessible. Thus, any phone with an Internet connection will be able to join the application server. And – I guess you can celebrate now." Johnny said, bursting into a full smile. "I think we just extended time into the past once more. Our app is now a reality." He said.

This time, it was Bill's time to get shocked; his mouth gaping open as he stared at the screen. "Are you for real?" He asked.

"I think the question you should ask first, is – who is going into the past first?" John shrugged.

"How long of an extension did we create?"

"Three days."

CHAPTER SIX

Present day

One day before the first victim disappeared.

Johnny remembered how happy they had both been a month ago, and now the very thing that brought them happiness was causing them immense pain and guilt at the moment. They were now sixteen years old, and although they never once proclaimed it to the world, they had both achieved a major breakthrough in science. An invention that was nothing like any other thing that had been invented before. They had created time itself.

On that very day, they had both launched their mobile app, which was the gateway to the past and the future. John could have sworn for accountability and attestation in the mobile cloud application. Both of them had made sure there was no mistake, but at the very end, something unforeseen had happened. The mobile cloud provider had stood frozen for approximately three seconds. And almost immediately, John and Bill had gotten to work, distributing and strategically placing virtual machines that monitors, analyze and regulate the mobile cloud application, and the associated content to detect and prevent malicious behaviors. After that, everything had proceeded smoothly, and the virtual machines continued to monitor all the virtual network application traffic, so as to draw conclusions in a distributed and autonomic fashion.

The beauty of the Virtual machines was that it included security features which can strengthen the mobile cloud application security, which include firewall, anomaly and intrusion detection, information disclosure analysis, anti-virus and taint checking. Those services, which are too computationally intensive to be executed at the mobile devices, can be easily performed in the cloud. In addition, the applications can be confined in a remote sandbox.

"I almost thought something went wrong." Bill had said, breathing normally once more. "Imagine if a mistake from us causes someone to be trapped in the past or the future."

"We are not Villains Bill. We never were, and we would never be." John replied softly. "And besides, there's a password algorithm we put in place. Only those that can answer those questions, can enjoy the benefit the app provides."

If only they knew.

CHAPTER SEVEN

Bill felt his heart sink as he watched the delete percent increase with each second. This was it. It was the end. Once the app completely deleted, it was really the end, and there was nothing they could do about it.

"Everything will go back to how it was." John said, crossing his arms across his chest, while pressing the back of his right hand against his mouth. His face seemed unaffected, but his heart was also breaking into a million pieces. He wanted to scream, reach forward and stop everything. *But could he?* The price to their dreams was too expensive. If they decide to ignore it all, many more people would get stuck in the void of time. Closing his eyes against the thoughts that tormented his mind, he could hear a few sniffles in the background, and he knew Bill was crying.

"I need a moment." His best friend said gruffly, followed by retreating footsteps as he left the room. Left alone with the silence, John opened his eyes once more, and the tears he was forcing down left his chest and throat burning. As he looked at the computer screen, he realized that the percent was now in ninety five. Five more, and the thing he was most proud of would be gone.

Looking down at his feet, images of all the times they had stayed up late into the night, to their first small success, and their biggest ones flashed through his mind. As he looked back up to check the percentage, he noticed something was awfully wrong. Reaching for the control, he tried inputting some commands, but nothing seemed to be working. The computer screen was filled with blue lines, and the algorithms he knew like the back of his hand began to overwrite, and cancel out the ones he and Bill had written. Johnny could feel the hairs on his arm rise, and sweat roll down his back. As he turned around to call his partner and best friend, the unthinkable happened.

Outside the room, Bill was resting against the wall, his head tipped backwards so he could look up at the ceiling. The rim of his eyes, and the tip of his nose were red. He wondered if this was how the parents with the missing children had felt; except theirs was living, and his wasn't. Breathing in deeply, he ran a hand across his face, and turned around to go back in. he wanted to be there while he watched the last moments of the thing he was most proud of.

As soon as he walked back in, it felt as if his heart fell to the ground, and a cold sweat descended over his body. Few centimeters away were one of his worst fears. The computer screen which had displayed the countdown was no more, instead, the screen was blank, and flickering light came up from time to time. But the worst realization was finding out his best friend was missing. There was only one answer to that. Johnny was also stuck in the void of time.

CHAPTER EIGHT

Have you ever been in a situation where all you felt was overwhelming fear? A moment when you feel immensely cold, and at the same time sweat rolled down your back?; a moment where your intestines felt like they were in a twist, and the tighter it became, the more it hurts?

Billy Martins was that person right now. He felt it all.

"John! Johnny!" He yelled, rushing out of the room; his eyes darting from side to side as he searched for his best friend. His mind knew; it knew the truth, alright. But he wasn't ready to accept it. "John!" He yelled once more, but this time his voice ended on a sob. "Oh God know… For God sake…" He groaned, his throat burning from the tears that were just beneath the surface.

"Bill…?" A voice called from behind, causing him to spin so fast, he almost had a whiplash. But it wasn't John; it was his fifty three year old father – Jordan. "Is everything alright buddy?"

"Dad…" Bill said brokenly, and at that moment, the tear he was holding on to fell. "John is gone." He moaned.

It had been ten years since Jordan saw his son cry. As he saw the tear roll down, he felt his heart stop. "Sweet Jesus…" He whispered softly, walking to where his son was standing. "Calm down, what's going on?" He asked, unable to keep the worry out of his voice.

"I was outside, I left him in the room, and by the time I went back in, he was gone. John is gone." Sixteen year old Bill cried.

"Perhaps he left for home… you know, perhaps he went home." Jordan said suggestively, placing a gentle hand on his son's shoulder.

"No, no… you don't understand. He never left. I was right outside the door. He never left, he just vanished. It is my entire fault, I should have never left. Everybody is gone. We should have never built the app. I don't even know what's going on." Bill rambled on, covering his face with both hands as he spoke.

"Calm down, calm down; what… what are you talking about?" Jordan asked, more confused than ever. He couldn't wrap his head around whatever he was saying, and nothing seemed to make sense.

Removing both hands from his face, Bill stared at his father; the rims of his eyes red. "The disappearances… the kids that keep vanishing... I had something to do with it." Bill confessed, unable to keep the truth any longer. Jordan didn't know whether to laugh or believe what Bill what saying. But one look at his son's eyes, and he realized there was something really wrong.

"You need to start from the beginning." He said instead, heading towards the office John had disappeared.

"No!" Bill yelled, tugging hard on his father's shirt. The force sent Jordan taking two steps backward; the collar chaffing his neck.

"What are you doing? For God sake Bill! We need to sit down and talk this through. I don't understand what is going on -"

"Don't go in..." Bill repeated, shaking his head. Deep in his heart, the fear that someone he cared for would go in and never come back, reared its ugly head in his mind. As he scanned Bill's face, he realized whatever was behind that door was causing immense fear in his son. *'What the heck is going on?'* he wondered.

"Let's go somewhere else... please don't go in." He pleaded with his dad once more.

"Fine." Jordan agreed. "But you have to tell me everything. Everything." He said pointedly. "You can't leave anything out."

Forty minutes later.

The tension in the air was so thick, a knife could cut through. It was forty minutes of Bill telling his dad all that happened. Forty minutes of uncomfortable silence from Jordan as he listened to it all. He couldn't believe his ears, neither could he believe that humans had the ability to transverse space and time; and most of all, he couldn't believe that his son was one of the people that invented it. "I swear to God if you're lying..."

"I swear dad. Everything I said is true, and now John is missing. I don't know what to do."

"My God..." Jordan whispered, running a hand through his face as he reprocessed everything he had just heard. He could feel the beginnings of a headache creep up his neck. His mind was currently blown, and the realization that his son's invention was the cause of the disappearance made him scared. "Who have you told? Who else knows what is going on?"

"Just you... John and I. You are the very first person I have talked to..." Bill paused, swallowing the spit that had pooled in his mouth. "You are the first person I have told."

"Okay... good...good... what... Uhm... okay, let me think." Jordan mumbled, rubbing his temples. "Uhm... what do we do, what do I do...." He moaned, unable to come up with anything.

Bill sat quietly, his thoughts blank, and his eyes dry. He didn't know what to think, or act or say at this point. Telling his dad had been his last straw; there was no way he could hide it anymore. "We need to get back into that room. I need to see this app, and see what I can do." He heard his father say.

"But dad –"

"We have to do something buddy, even if it involves going back to that room. Because, every moment we spend doing nothing, every kid stuck there could be going through something worst." At his father's words, Bill felt a twinge in his heart. "But first, I need to do something. I need to call Maria."

Hearing the name of John's mom, Bill could feel his spine stiffen, and once again he felt his palms grow clammy. "Are you going to tell her...? Tell her that John disappeared?"

"No... not yet. Not yet..." Jordan repeated, dialing her number.

CHAPTER NINE

"Ready?" Jordan asked. Before Bill could say anything, he turned the knob and pushed the door backwards with more force than he wanted. For some reason, as soon as he opened the door, he expected something to pop up and attack. The whole moment felt very bizarre to him. But unlike his imaginations, everything seemed normal, very normal. Stepping in slowly, he looked around, and then at the computer. Bill was only a step behind, and as he stared at the computer, he knew he had developed a trauma due to the whole incidence. The computer seemed to be in hibernation, and as he watched his dad touch the mouse, his heart banged with a pang.

"The computer is dead. Did you unplug the power cable before you left?" He asked, pressing random keys on the keyboard.

"No, the computer was switched on all along. The app was supposed to be deleting; it was deleting before I left. I don't think he switched it off." Bill explained.

Reaching down, Jordan checked to see if the plug was loose, but it was okay. "What is going on…?" He muttered, switching off the socket, and then switching it back on; but there was no difference. Frustrated, he pressed some random keys again, and this time the computer switched on. "It's on. It's on!" he repeated, unable to keep the relief away from his voice. But it was only short lived. The screen suddenly went blank once more, and then turned blue. Almost immediately, codes began to appear at a fast pace, and different lines began to change. "Bill… what's going on? I don't understand what's going on. Can you do something?"

"I can't!"

"What does that mean?"

"This is not us…This is different." Bill said, biting his fingernails.

"What do you mean?"

"The codes are different, we didn't write that. The algorithm, the SPL and GC34 it is all wrong. After adding the call to Get Loge to the end of the sqlmap main application, so we can iterate over the log messages and print them to the screen v for us to see when the scan is finished. And also, there was a scan code on the full sqlmap to retrieve the results, by bypassing the Bad Store URL as an alias to the application. But none of that is showing. The request to sqlmap is all wrong. The results should look something like listing 9-29"

Jordan was confused, and everything his son said made no sense. "Hold up." He said, putting a palm up. "Does this mean you were hacked?"

"No, no… everything is the same, but the codes are all switched… it is almost like the app has a mind of its own; which shouldn't be possible at all."

Sighing, Jordan straightened up. He was having a full blown headache now. Picking his phone up, he knew there was a number he needed to call. Almost immediately, the voice of an older male filled his ear. *"Hello? Is that you Jordan?"*

"Yes...It is me. Long time no see."

"It has indeed been a long time... twenty years too long. How are you?"

"Yeah... yeah..." Jordan cleared his throat. "Uhm... about that..."

"Is something wrong?" The man on the other side of the phone asked.

"Yes. Code orange."

"I will send you an address."

"Okay... we will be there soon.... And..." he trailed off

"Yea?"

"Thanks."

Four hours, fifteen minutes and seven seconds. That was the amount of time it took for Bill and his Dad to drive down to Charlotte. The wind coming through the windows was their company, and the sound of the tires tearing the road was their song. All through the journey, no one said a word. It was not an uncomfortable silence; but a silence filled with thoughts and unsaid words. Finally, Jordan slowed down as he turned into a new lane. This time, it was a residential area, and houses lined each side of the street. Backing into the driveway, Jordan killed the engines.

"We are here…" He said, removing his seat belt. Bill looked up at the house few centimeters away, and he couldn't help but feel nervous. Their echoing footfalls felt much louder in his head as they walked to the door. As soon as Jordan rang the doorbell, Bill could hear the blood rushing through his veins. Soon enough, they could hear some noise coming from inside the house, followed by hurried footsteps. As the door flung open, it revealed a man who looked similar in age to his father. He was wearing a black Lakers vest, with a face cap that read paws. "Jordan!" The man yelled, laughing as soon as he spotted them.

"Marcus!" Jordan greeted back, hugging his longtime friend, and college roommate. "It is so good to see you." He chuckled, pulling back. Jordan felt happy to see his friend, and the rush of memories made him feel nostalgic. But most importantly, he knew the man in front of them was their last hope.

"You look the same." His friend sniggered.

"Nah, my bones creak these days." Jordan replied, chucking. "I see you grew a moustache."

"This is my son, Bill." He introduced, making way for Bill who was hovering at the back.

"Ah, he looks just like Laura." Marcus guffawed, hugging Bill. "Nice to meet you son."

"Ah, ha… Thank you sir." Bill replied awkwardly, hoping the greetings would be over soon.

"Come in, come in." Marcus ushered, opening the door wide.

As soon as Bill passed over the threshold, his eyes scanned the new environment he had just stepped in. The interior of the house was nothing like he had imagined. When his dad told him they were going to see someone in the tech business; he had imagine… tech everywhere. But what he was seeing was a normal home. "Come through here." Marcus guided, leading towards a room at the end of the hallway. The room looked like a study; with the typical, desk, chair, and bookcases lining both sides.

"Here we are."

"Thanks man… I really appreciate you seeing us on such short notice."

"Old age made you so formal uh." Marcus scoffed. Walking towards the opposite wall, he placed his both palms on it, and began to feel around. Bill wondered what he was up to, and in the blink of an eye; the wall opened up to a new world. As soon as the doors opened, Bill retracted his earlier thoughts. This man was built different. "Wow…" He muttered.

"I know." Marcus grinned, proud of his workspace. "I built these five years ago. And there is always a new addition every year." Taking them to the far corner, Marcus showed them another small office by the side. "We can talk here." He said. As soon as they took their seats, the air in the room changed almost immediately, and the earlier light-heartedness seemed nonexistent.

“Code orange?” Marcus said.

“Yes… but – It’s my son.”

CHAPTER TEN

Marcus turned towards Bill, his brows raised. "Your son?" He repeated. "What happened?" He asked. Bill looked at his dad's face, hesitant about telling a total stranger everything. "It is alright son… you can trust him." Jordan said, patting him on the shoulder. Breathing in deeply, Bill turned towards Marcus and said. "I built a mobile app with my friend."

"Yes…?" Marcus said, urging him to proceed.

"But, it is not just any app; but an interface that is the key to time." Bill explained. "In simpler terms, we built a time machine." He finally said. An impregnable silence filled the room, and as Bill looked at the man he had bared his soul to, he saw an expression he was expecting.

"That's it?" Marcus asked. Before Bill could say another word, he began to laugh; a boisterous laugh that came from deep within. "Oh my God Jordan, you got me this time. You really did get me." He chuckled. "I should have known you would be up to this, and your expression, oh my God; I totally believed everything."

But Bill and his dad were not laughing. This was a life and death situation that left their insides turned upside down. "It is not a joke Marcus."

"Yeah right." He snickered while slapping his knee. "You really got me this time."

Bill looked at his dad, wondering if this was going anywhere. "Dad…?"

"Show him." Jordan sighed.

"Show me what?" Marcus asked, looking from Bill to his dad, then back to Bill.

"Give me an untraceable laptop."

"Are you for real right now?" Marcus asked, looking at Jordan this time.

"I am."

It only took few minutes for Marcus to produce a laptop. This time Bill was the only one seating, and both his dad and Marcus were standing behind him; so they could see the screen too. It was like an invisible pressure to Bill, but he knew he had to make Marcus believe him. Removing the hardware from his bag, he plugged it in, and immediately got to work. As soon as his hands got to the keys, it felt like he was in his natural zone; a habitat specially designed for him, and almost immediately, he forgot above everyone and displayed the skills he had. He began to create a copy of the codes that had started the whole thing. The screen began to fill with codes; a beautiful array of numbers and letters, and sometimes special characters. It was a semblance of the future; an intricate construction that left a deep impression. Once again, the screens started to fill with lines of different color codes, flashing-up and racing up on the screens, and in the background Athena' thousands of blue server lights, flashed like blue stars in the night... all the display screens go black, and then a white cursor just flickered and flickered, and then the screens, changed into IT blue, and fill with racing lines of white bios code.

"Sweet Maria…" A voice appeared so close to Bill's ear, he had no choice to look sideways. It was Marcus peering over his shoulder, with his face few inches away from his. "You created this?" he asked.

"Yes… my friend and I did."

"How did you sequence the algorithm to be 6'4 interface of the eighty sixth formats?"

"Code CCPP is the extent of the of the RR25 code, we could bypass the Ender-one file disturbance to another directory information pathway." Bill said, his brows drawing close as he stared at the code lines on the screen. "The hook libraries make a compelling skeleton for anyone wanting to write cross-platform tools quickly and easily. Running the code from Listing 5-5 should decode both the string "hexagonal octal"

"What of the CCPD error… you didn't encounter that while you were writing this?"

"The algorithm Apollo helped to bring out from the information bank. Athena data risk embraced the open source software in ways many people would have thought unimaginable; and with the help of AlPine60 and Beta70.5, we were able to cancel out the CCPD error."

“Oh my God…” Marcus whispered, followed by a long line of cusses as he straightened his frame. “This is just… didn’t you loop the recon to the Sqlmap?”

“The Test Post Request with Sqlmap method accepts four influences. The first influence is the URL that will be sent to sqlmap. The second influence is the data that will be in the post body of the HTTP request—be it POST parameters or SOAP XML. The third influence is the value that will be passed in the SOAP Action header in the HTTP request. The last influence is the unique value that is vulnerable. It will be replaced with an asterisk in the data from the second influence before being sent to sqlmap to fuzz. After we bypass the message to the screen we might be able to get a contemporary URL; and once we tested it out, it worked. So we created a Sqlmap Session and Sqlmap Manager objects.”

“I need to talk to you.” He said, turning towards Jordan. Pulling him to the side, he whispered. “Is there something else you need to tell me?”

“There’s –”

“You need to tell me the truth.”

Closing his eyes for a split second, Jordan finally confessed. “The disappearances… they are linked to this.”

“What dis –” Marcus trailed off as something clicked in his mind. “The vanishing case… All ten of them…?” Marcus asked, surprised.

“Now eleven… the last one hasn’t been reported.”

Marcus eyes widened more as he stared at his friend. He had once believed nothing could make him rethink about life; he didn't believe that anymore. "There are two things I need to let you know." Hearing those words made Jordan anxious. He tried bracing himself for the worst, but no matter how mentality he prepared, he knew he was not ready. The words 'there is nothing we can do' crossed his mind', and he could feel his hands tighten into hard fists. 'What would he do if that was the case?' he wondered.

"What...?" He whispered back, finally saying something.

"Your son is... if I could describe him... Your son is a mad genius. And, we need to hurry, or it will be too late. Looking at the algorithm pattern, we have twenty four hours before everything goes haywire. And by haywire, I mean before it becomes an APP pandemic. It could affect the whole world."

CHAPTER ELEVEN

Marcus gets to work immediately, with Bill by his side as they try to decode the overwritten codes that were persistent. Their screen is continuously filled with ever-increasing code lines, and looping tabs that cropped up almost every second.

"It has gotten to a point whereby I am not surprised anymore. Do you see this? It is linked to 'y86 assembly… how the hell did an app gain a mind. This is one of those minds blowing, fucking unbelievable kind of moment. You guys created something this crazy, and you are only sixteen? Fucking hell..." Marcus cussed, shaking his head as he stared at the screen.

In any other situation, Bill would have been incredibly proud as he showed off his creation, but at the moment, he found no joy in what he created. "We created something evil." He mumbled, but Marcus heard his words.

"Sometimes things go wrong, and this is one of those moments. We are going to fix this son. We will." He assured, jutting his chin towards the screen.

Fueled by the courage offered to him, Bill continued to unravel the codes that seemed to attack them. The fear that one wrong move could make them disappear hovered in his mind; but fighting for the life of the one he cared for, and for those who had someone waiting for them, kept him going on. "If only we had a stronger link… like a satellite link." He mumbled once more.

"Why?" Marcus asked.

"Hmm…" Bill asked, looking up.

"You said satellite link. Is the usage of the satellite the only way we can gain access to the data risk?"

At that moment, Bill thought. *'This man has good hearing.'* "I don't know for sure, but it might increase our chances of beating this."

Marcus stared at the string codes he just entered, then back at Bill. "Did I tell you before? You are a genius, you know that? The communication satellite operates at frequencies near four gigahertz for down-linking, and six GHz for up-linking. If we are able to access it, it would give us a stronger signal. Wait… let me think… how can I access the LEO…"

"LEO?"

"A Low Earth Orbit satellite. It has a lower coverage area than MEO satellite; but the iridium system has sixty six satellites. It is also capable of storing data received while passing over one part of earth and transmitting it later while passing over another part; like the CASCADE system."

"Canada's CASSIOPE?"

"Yes! I just remembered I have a buddy back there. I will need to gain permission for the access code we would need. I will be back in few seconds." Marcus said, standing up and leaving the room. As he walked down the hallway, he thought about the young boy his friend had given birth to. Bill Martin. His father was a man in his early fifties, with eyes that showed wits, and a personality that portrayed his charisma; it wasn't hard to imagine how Bill would look when he got older since they looked so much alike. The major difference was how much of a genius his son was.

"Are you guys done?" Jordan asked; standing up from the chair in the study.

"No, I want to make some few calls. But there's progress." Marcus watched relief wash over his friend's face, and he couldn't help but smile. "We will be fine." He said, nodding his head as he left. Marcus came back few minutes later to the room, relieved he had gotten the code "Anything?" He asked, drawing Bill's attention.

"No… just building a hole for the code will need." He replied. "Did you get access to it?" He asked.

"Yes… but, it would take twenty minutes before it gets back to us." Marcus sighed, sitting back on his chair. His screen still had the LEAR bypass, causing a furrow to appear between his brows. "Never seen that type of assembly before..."

"It's linked to 'x86 assembly… an old game code." Bill explained. "My friend and I thought it would be fun to add it."

"What else did you add?"

“A bunch of personalized questions that serves as a key to the main server whenever anyone launches the app from their phone”

“What kind of questions?”

“Just random questions and stuff… Like the secret tunnel that most teens in our community knew, like what line did Lei say to Erin the first time they met, the nickname of adrenal Test, and some other’s.”

“Adrenal Test? Who is Levi?” Marcus asked, puzzled.”

“Lei… from attack of the beasts? Does it ring a bell?” Bill asked, turning to look at Marcus.

“No, never heard it before in my life.” Marcus replies, feeling the generation gap between them for the first time. The screen of his phone lit up, catching his attention. As he unlocked his phone, Marcus saw the code they had been waiting for. “Yes!” He said excitedly, jumping up from his seat. “I got it! I got it!” He laughed, showing his phone screen to Bill.

“The satellite link?”

“Yes. I finally got a safe directory to the link.” He explained enthusiastically.

“You have a new message.” Bill said, pointing to his phone

“Oh, yeah… thanks.” Marcus replied, turning the screen back to himself. His smiling face changed almost immediately, leaving behind a scowl. “I need to ask you something Bill.” He said, lowering the phone in his hand.

“Yeah?”

“The database risk used for the mobile app. What configuration interface did u use?”

“IPM 65. On the computer back home.”

“Shit!” Marcus cussed, gritting his teeth. “We have a problem.” He groaned.

Bill froze for a second. *Did someone else disappear?* He thought. The blood in his face began to leech, and he was as pale as snow in the blink of an eye.

“It is the link.” Marcus said, causing Bill to gasp. It felt like he had been holding on to his breath from the start, and he realized he would have not been able to carry on if another person disappeared. Blood began to rush back to his head, and he felt faint. “What about the link?” He said in a low voice, resting his arm against the table as he spoke.

“We need to input it directly to the home computer for it to work; the sad news is the link will be expiring in two hours, thirty minutes.” Marcus said, grabbing some things off the table.

“It took four hours, fifteen minutes for us to get here.” Bill added, feeling his vision get even blurrier. *Was life finally giving up on him?* He wondered, as the feeling of hopelessness began to set in once more.

“Let’s take my car.” Jordan said, appearing from the doorway.

“We have two hours, thirty minutes to launch the code from the home base. How fast can you drive?” Marcus asked, packing some other things into his bag.

“Fast enough to get you there.” Jordan said, leading the way. It was race against time.

CHAPTER TWELVE

Jordan takes the driver seat, while Marcus sits next to him; leaving Bill to sit behind. Breathing in deeply, Jordan hastily adjusts the side mirrors, as he mentally prepares himself for the crazy drive ahead. Pushing his right foot on the left hand pedal, Kim pressed on the automatic button; the soft purr of the breathing engine filling the air.

"Sweet ride…" Marcus said, eyeing the steering wheel.

Jordan looked at him from the corner of her eye, wondering if he was the only one that was enjoying the whole situation. "Everyone hold tight." He muttered, revving the engines and spitting granite as the wheels forcefully pushed against the ground. As he drove out of the driveway, he reversed sharply to the left, lurching them all to the side.

"Hold on…" Marcus said, holding on to his temples. "I remember… I remember now. Jordan Martins, the maniac driver. How could I forget?" Marcus laughed incredulously, turning towards Jordan. "How could I forget? Are we safe?"

"A maniac driver is what we need right now." Jordan replied, not taking his eyes off the road. "I will all be fine." He assured, as he increased the speed of the car; a complete contrast to his words as she lurched them all forward. As they hit the freeway, the nerve wracking moment began once more. Jordan swerved the car to the left against a sharp corner, the impact of the move causing them to momentarily jump up their seat. Marcus wasn't so lucky, and in the process, hit his head on the roof of the car. And the crazy driving didn't end there.

To shave off time, Jordan didn't remove his leg from the gas pedal at all. As a car swerved towards their direction, Jordan changed gears, rolling the steering wheel to the other side. The sharp turn, tipped the balance of the car precariously close to the side; leaving the car almost off the road. As the car screech loudly on the road, dust pulling up in the air as the smell of burnt tires is all around, he kills the ignition. "We are home." He said, getting out of the car. They made it back in one hour, twenty five minutes.

Bill looked at his dad, wondering who that person was. It was like he made out alive from an episode of Furious five. Still shaking, both passengers managed to get out of the car, and miraculously set up the intra code in time. Just like Bill predicted, the LEO link was their best shot on overpowering the APP's contra.

"With the LEO connection, we can overwrite the codes, and take it back to normal."

Bill didn't show any emotion as he heard those words. He was scared once he showed he was happy, life would take it away from him and something bad would happen. He was scared everything was a joke, and there was nothing they could do about it. "All that's left is for us to delete it."

"What?" Marcus asked, looking at Bill like he was crazy.

"I said, all that's left is for us to delete –"

"I heard you clearly. But why would you delete it? Why would you want to do that?"

"To return everything back to how it was."

"Is that what your friend would want?"

"John would want nothing more than to be back home."

"Once we shut this down, every moment in time when THE APP existed will vanish." Marcus said, folding his arms against his chest. "Is that what you want?"

"I know that." Billy replied, looking down at his feet. The silence that followed made his decision weigh heavily on his mind. But he knew he would rather not have the app than not have his best friend, or save the rest of the people that disappeared.

"You know…" Marcus started once more, his voice breaking through Bill's thoughts. "You could decide to be selfish, forget everyone and proclaim to the world that you invented the greatest thing on earth. Imagine the amount of money and fame you would get. Just think about it!" He said excitedly, grabbing on to Bill's shoulders. "Imagine how much you will be contributing to science. The whole world will be at your feet and you can do whatever you want, you can –"

"Marcus, enough!" Jordan yelled, his voice booming across the room. "You need to stop." He said through gritted teeth, moving closer to where they stood. Bill felt scared, and as Marcus removed his hands from his shoulders, he couldn't help the unsettling feeling that appeared in his mind.

"I… I am sorry." Marcus said to Jordan, running a hand through his hair. "I am sorry." He repeated, turning back towards Bill. "I must have gone crazy for a moment there." He groaned, turning his back towards the computer screen. "Do what is right." He muttered.

Breathing in deeply, Bill walked slowly towards the computer, and with a heavy heart, he began the deletion process.

"Do you wish to delete this database completely?" The notification read.

Looking up, Bill blinked rapidly, trying to stop the tears. "Yes." He clicked. This time, he watched from the beginning to the end as the app and all the database they had built over the years vanish. He felt no different as the percent reached hundred, neither did he disappear. He had expected something huge to come out of it, yet once again – everything seemed normal.

He stepped three steps backward, then turned around to run out the room. As soon as he opened the door, he bumped into someone that brought tears to his eyes.

"Why are you crying? Did something happen?" John asked

"Nothing… nothing happens." He replied, sniffling as he smiled.

CHAPTER THIRTEEN

The distant sound of people playing basketball was carried by the wind; a familiar and serene sound. The world was at peace. The vanished children returned, and the world never knew they vanished. It was like the pandemonium had never happened.

"God… I can't believe we have math class right after physics." John grumbled, unwrapping the burger he had just gotten.

"It could be worse; you could have chemistry after both physics and math."

"Yeah, yeah… I wonder why your timetable is so good this semester. Did you bribe the administrator or something?" John asked, raising his head. "Why are you looking at me that way? No way… did you really bribe the administrator?" John asked incredulously, a bit shocked. "How did you do that?"

Shaking his head in dismay, Bill unwrap his own burger. "You idiot." He mumbled, biting into it. *'He doesn't know…'* Bill mumbled, shaking his head. No one knew anything. It was like every single moment the app existed, alongside the memories about it, simply vanished. How come his own memories were left intact? Bill wondered. He had so many questions, and yet – there was no one or no way to answer them all.

"Oh shit!" John coughed, as he choked on his burger; his voice breaking into Bill's thoughts.

"What?" Bill asked, raising his head up.

"Uhm... four O clock, now." John whispered back, cleaning the corners of his mouth as he spoke."

Bill was a bit confused, unaware of what was going on. Clearing his throat, he leaned backwards, pressing a half fist against his mouth. Slowly, he turned towards the direction John had mentioned, looking for anything that could have stood out. "What... what?" He repeated, wondering if the person or thing had left.

"Dudeee, come on! I said four O'clock... the... the girls, okay. You see them!" John explained, talking through gritted teeth.

"Oh..." Bill muttered, finally spotting the people John was talking about them. "What about them?" Bill asked, trying to see if they were familiar.

"The girl with the dark hair? She is in my chemistry and biology class."

"Is she?" Bill smirked, turning to look at his friend. "I see you noticed." He said, barely keeping the grin out of his voice.

"Shut up!" John grumbled, wrapping his remaining burger.

"So what's her name? Have you talked to her?"

"I don't know..." John shrugged, wondering if the tip of his ears were as red as he thought them, to be.

“Okay….” Bill replied, standing up.

“Where are you going?” John asked, looking up. At that moment, the intensity of the sunlight increased, and a patch of light hit him in the eye, causing him to squint. “Where are you going?” He repeated, blinking fast to clear his blurry vision.

“We have a name and a number to ask.”

“What?” John laughed, not believing his ears. “What the hell are you doing?”

“Do you want her name and her number?”

“Yeah but…” John trailed off as he stared at his best friend’s face. “No you don’t!” He said in a low voice, instant regret washing over him. He knew what Bill was about to do, and he couldn’t stop the nervous sensation that filled his belly.

“Come on.” Bill shrugged, about to walk off towards their direction.

“Bill, no! Yo… Bill!” John whisper yelled, slamming his left heel on the ground. “Bill, if you leave I am never talking to you.” John threatened, mortification written on his face.

Bill turned to look at bill, a mischievous smile on his face. “Trust me bro, you don’t want to waste another moment if you knew what life could throw at you in any minute.” He said, running off towards the girls.

John lifted his bag, wondering if he should run away before Bill came back. His friend had been acting crazy from a week ago, and John was beginning to wonder what exactly happened to him. Almost as if he heard his name, Bill turned towards him, pointing at his direction. John felt like falling over as the girls turned to look at him also.

"I am killing you Bill, I am going to kill you." He mumbled, wondering if he should still leave at that point.

"John, come over!" Bill yelled from the distance, waving his hands, and beckoning him to come. This was the second time that day that John regretted it all. He wished he had never said anything. And as he slowly walked towards them, he thought about things to say, or do… or act.

"You are here." Bill smiled, patting him on the back. "Ladies, meet my friend, real cool dude." Bill introduced. "He is pretty good at sports too." John wished Bill could just shut up, and as he stared at the girls he was talking to, he realized he couldn't meet their eyes.

"I am John… Johnny. You can call me John or Johnny. Doesn't matter." He laughed nervously, turning to look at Bill.

"Hi, I am Rita and this is my friend Alicia." One of the girls introduced back, causing John to finally meet her gaze. It was the black haired girl he had a crush on, and as he looked at her, he couldn't help but smile inwardly. *'So her name was Rita…?' He* thought.

Rita was a five ft. five, Caucasian; with grey eyes and a beautiful smile. John couldn't help but think she was extremely pretty, and although her friend – Alicia was pretty too; but in his eyes, she fell short.

"Yeah... nice meeting you." He smiled once more, wondering if he looked like an idiot at that moment.

"Bill told me you are in the same Biology class as I am. I was wondering if you would like to join our group project." She asked.

"Biology... Sure. Why not?" He responded, smiling again. "I mean..., if you don't mind."

Bill watched from the sidelines like a proud parent, and although he wanted to laugh so badly, he knew it would ruin the moment. As he looked up, he realized the friend – Alicia was also leaving them to the moment. Giving an acknowledged nod, he watched as she returned it. They both spoke the language of friendship, and just like that – John got the number, and the name of the girl he had a crush on.

CHAPTER FOURTEEN

"Have you been listening to me?" Bill asked, raising a brow as he stared at his friend. "Huh… Sure." John responded, his eyes still glued to the screen of his phone.

"Come on, are we still playing or not?" Bill complained, throwing his hand up in the air. "What the hell are you doing, grinning and looking at your phone all the time?"

'Just replying some… uh…" John looked up briefly, catching Bill's irritated expression. "I was just uhm… checking the assignment Mrs. Wright gave us the other day."

"You have been grinning to your phone, and you are telling me it is because of your assignment? Who are you trying to fool?"

"Alright, Fine. It is Rita." John sighed, finally confessing.

"Oh…" The expression on Bill's face changed almost immediately, and his annoyed expression changed into a grin. "Look who has been texting his girlfriend."

"Come on, she's not my girl…" He trailed off, looking at his phone, "yet." He completed.

"Oh ho! Look who we have here. Mr. Confident, I am going to get a girlfriend soon!" Bill teased, wiggling his brows.

"Shut up! We were chatting, and she told me she built an app. Can you believe that? I never knew she wasn't into that kind of stuff." John said proudly, eliciting a disgusted look on Bill's face.

"What?" John asked.

"All I see right now is a whipped man."

"Get out of here." John mused, throwing a book towards Bill.

Catching the book, Bill laughed. He was glad to see his friend happy, and he couldn't help but tease him about that.

"So, I invited her over, and I told her I was going to show her the app we both built. I hope you don't mind?"

"What…?" Bill whispered, dropping the book to the ground. "What did you say?"

"The app? The one we both built… I told her to come over so I can show it to her." John repeated, shrugging.

"You remember? When, Since when!"

"Since forever? Why are you asking me that…?" John asked, tilting his head to the side.

"Answer me!" Bill yelled.

“Okay… alright…” John laughed tersely, dropping his phone. “What are you talking about? The assignment finder app we built two months ago… you forgot that?”

“Assignment finder?”

“Yeah… Why are you yelling all of a sudden? That scared me.” John grumbled, picking his phone back up. “But Rita is still coming over, you cool or what?”

“Yeah… yeah… she can come.” Bill replied absently, still disturbed. He could still feel his hands shaking, and he knew if he stayed any longer. John would find out something was wrong. “I need to uhm… check something.” He said, not able to think of another excuse. Leaving the room immediately, he walked fast until he was outside; in the open skies.

As the wind caressed his face and ruffled his hair, he tried to calm his thoughts. There was no way John could remember. No way at all. He had not mention a word, and Bill knew he would forever keep it that way.

Minutes turned into hours, and hours turned into days, until finally Three weeks passed. Bill’s outburst was now a thing of the past, and he could keep his expressions better controlled whenever he heard John say anything about an app.

“I am leaving for work!” Jordan yelled from the doorway, causing his soon to pop his head out of the room.

“Will you back for Dinner?” He asked.

“Yep. I can make it back in time.” His dad

“Fast and furious.” Bill chuckled, causing a crease to appear between his father’s brows.

“Huh?” Jordan asked.

“Nah… it is nothing.”

“Okay, I will see you later.” Jordan replied, opening the door.

“Dad, wait…” Bill called out to him once more, causing him to turn around.

“Anything else?”

“I… I just wanted to say thanks.” Bill said, feeling a bit shy.

“For what?” Jordan asked, raising a brow. “Do you need money?”

“No… well, if you wanna give me. But, I just wanted to say thanks. For everything.” Bill replied, walking away before his dad could say something else.

Jordan felt warmth spreading through his chest as he watched his soon escape. He didn’t know what he was thanking him for, but he felt good. Whistling happily, he closed the door to the house.

CHAPTER FIFTEEN

Outside of the police station, the sun stood gloriously above, shining brightly with its golden rays tinting the light blue sky that spanned as far as the eye could see; a very good day if one could say. Though the weather was great, it did nothing to reflect the seriousness that went on inside the station. In one of the rooms, the mood was dark, and the messy appearance of strewn papers, case files, and the LED lighting did nothing to improve the mood, or the tension that filled the air-space. Detective Key stared at the four pictures sprawled out in front of him, his eyes darting back and forth between them. No matter how much he stared, nothing seemed to make sense at all. Nothing at all.

"This is not a dream.... It is happening once more..."

"Still at it?" The familiar voice of his partner drew his attention, causing him to turn around.

"You said the same thing... This is beyond vertigo. You freaking said the same thing!" He stared incredulously, his mouth gaping as he stared at his partner.

"What in hell are you talking about?" His partner frowned. "You definitely need coffee at this point. Thank God I brought you one."

“Stop playing with me…” Key whispered, looking at the coffee, then back at his partner. “Stop playing with me.” He said, but this time louder.

“What are you talking about?” Max chuckled, scratching his nape. “Are you that grateful I brought coffee? Anyways, I just got back from my rounds. I also learnt nothing. No one saw her that day. According to her mother, after the outage, she went up to her daughter’s room to check if she was okay, but she was not there. She first thought she had left her room to look for her, but no matter where she searched, her daughter was nowhere to be found.” Maxwell responded, staring down at the pictures on the table. “It is like every one of them vanished.”

“Okay, I am done.” Key cussed, dropping the coffee on the table. “It is either I went into the future, or somehow drugs got into my system.”

“What are you saying?”

“This –” Key paused, stretching his hand out. “Every single thing, every single word! Everything! And I mean, everything is the same, how is that fucking possible!” He snapped, banging his hand on the table.

“What the hell is going on with you? Did you really take drugs?”

“The technical analyst is going to call now.” As soon as Key said those words, his phone began to ring; and on the screen was the name of the analyst. “You see what I am saying? This is some high tech voodoo shit! It is like I am living the same thing all over. Can you believe this?”

"Okay, hold on Doctor strange!" Max replied, holding a hand up. "You do know that your joke is no longer funny. Of course we both know he was going to call by three; and so… he is a little early, how is that strange?"

"Okay, alright, you don't believe me uh. Just you wait!" Key fumed, putting the phone on loud speaker, before dropping it on the table. "You have good news and bad news, don't you?" He asked

"Well, yeah?" The voice on the phone responded. "Did someone from the lab call you?" He asked

Key looked at his partner's face, and then back at his phone. "No, no one did. But I know what the good news is." He replied.

"For God sake, Key what are you doing?" Max grumbled. "Let the man talk."

But Key wasn't listening; his mind was on proving he was not crazy. "You were able to connect three interfaces, weren't you?"

"Well, yes… I was able to connect three major interfaces from the phones retrieved from the crime scenes. Reaching DNS, and using Arduino IDE, Linux-based tool and mirroring pHreaker code hack, I was able to access the VPN radio network, using digital modulation. Three modulations had its IP address hidden from all incoming traffic and any networks..."He explained. "But that's not all, AI advanced global sniffing algorithm code was mixed with the retrieved algorithms '*Compaq UFI*, to walk through or you could say… map out the pattern directory."

“But you were not able to get it, right?” Key asked, looking at his partner’s face once more.

“Did someone from the lab call you?” The analyst repeated. “But there is no way that is possible…. I mean, just seconds ago we were able to create a new task and set the current options x. while paying special attention to the data option y. This is where we replace the vulnerable value in the post data with an asterisk. The asterisk is a special notation in wall-map that says we still need to set one more option before we can start the task. We need to set the correct content type and SOAP action in the HTTP headers in the request. Otherwise, the server will just return with 500 errors. This is what the next part of the method does, as detailed in listing 9-33.”

“Max doesn’t understand what you are saying.” Key completed before Max could say anything. “He wants you to speak English.”

“How –” Max started, but the voice on the phone cut him short a second time.

“Sorry about that. What I meant was, there is software I used which works as a reverse-engineered ethical hacking code to breach through, and pick out similar… Uhm… similar things between all the phones we brought into the lab. Each and every one of them had an almost identical LEAR, except their IP address was different, but it is definitely connected.”

“But even though you were able to connect the interfaces, you were led to nothing. We know it’s the same person, but we don’t know who and why, correct?” Key asked.

“Well yes…”

“And you are not sure if it’s a cyber crime since the modulation is being read from phones, and it is more like a complicated nothing?”

“Well yes…”

“Alright, thank you. I will get back to you later.” Key said, cutting the call.

“What was that?” Max asked, spreading his hands out in confusion.

“I don’t know.” Key replied, folding his hands against his chest. He now had his answer, but it wasn’t one he wanted – he could say, he had gone crazy.

Little did he know that he was the second person to retain his memories from the time, time vanished. He was completely in the dark

---

But max and Key weren’t the only one confused. As soon as the call ended, the analyst stared at the phone in his hand, not sure about what had happened.

“Is everything alright?” Grace, the second lab analyst asked.

“I don’t know.” He replied, placing his phone on the table. “But I just had the weirdest phone call ever. Did someone from the lab call the station?”

“No.” Grace said, shaking her head. “Did they say someone did?” She asked

"No… but." Before he could complete his words, the phone rang once more, and on the screen, he could see an unknown number.

"Another person from the station?"

"I don't know." He replied, placing the phone against his ear.

CHAPTER SIXTEEN

Feeling cold is a perception of decreased body temperature or the feeling that your body is colder than usual. It happens most times when your body temperature is below normal. For Bill, it was due to fear; the fear of the known that threatened to overpower his senses. The chills continue to permeate his bones, lifting the hairs off his arm. As he stared at the TV, he realized the same thing was happening, but some things had changed. He was living out a scene he could clearly remember, but this time it was him, not John.

"At around twelve forty five P.M., it was confirmed that seventeen year old male, Peter Langston vanished from his home, and has yet to be seen since then."

"The name... the name is different... the person is different...." He mumbled, pandemonium running through his mind.

"Police are ramping up the search for seventeen year old Peter, who is now missing for one week. According to eye witness, his mother,"

Just as the news reporter mentioned those words, the TV began to glitch, and the lights went out for a second. "It is starting over... this nightmare! It is never leaving! It is never going to end!" Bill muttered frantically, holding to the sides of his head as he paced the room. Reality was now like a movie, and the blurred lines between sane and insanity was thinning out. For Bill, it looked like he was hallucinating; a nightmare that bleached out the colors of his world. It was a horror. A horror only he could remember.

As if to confirm those words, the TV went back on, and this time, the words were different once more. It was back to the former.

"Laura Linn; said she never left home, and was in her room all through the time. Concern for Piper is continuing to grow as it has now been a full week since she was last seen. Detective Inspector Garrett Cornwall said "Enquiries are continuing in our efforts to locate Piper. And that there is possibility that her disappearance could be linked with the nine other cases of vanished people that happened within the month. The signs left behind are indicating so; which includes, a flashing mobile phone, and power outage in the vicinity of the crime scene. He is asking members of the public to come forward with any clues they might have. He also stated that, the police departments were supporting friends and family of the missing people, and is appealing to the missing people to contact home so that they can know that they are safe and well. Friends and family face an anxious wait for her return, or to find out what has happened to her."

The news reporter paused for a few seconds, and the tension in the room thickened further. It didn't seem possible, but it felt like the oxygen in the room was being sucked out "We just got new report that almost fifty thousand people have vanished worldwide; in a second." The reported announced, her voice quivering on the last words. "Please, if you are watching this, stay at home. The numbers of those missing are only those accounted for; they are not sure how many more is currently missing. Directives from the president is that _"

“What are you doing?” A voice from behind broke through the fog surrounding Bill, causing him to spin around so fast, he almost had a whiplash. His face was already ashen; and sweat had matted the hair around his temples. He couldn’t breathe, and his vision was getting blurrier by the second. “What…” He said almost inaudible, disconcerted and confused.

“I asked what you were doing.” His best friend asked. “What’s wrong?” He asked again, noticing the pale palette of his best friend. “Are you alright?”

“It started all over. It’s worst… It started.” He began to gasp, holding on to his chest as he met the floor.

“Bill!” John yelled, running towards where his friend had collapsed. “Bill! Mr. Martins! Mr. Martins!” John yells, hoping Bill’s dad could hear him.

“We need to do something… We need...” Bill continued to gasp, and he could see five Johns as he stared at his friend.

“Stop talking. Try to breathe, I need to get help.” John responds, trying to get up, so he could get his phone.

“You can’t leave… You can’t. If you go, you might also vanish. The whole world is vanishing and I don’t know how to stop it.” Bill cried, holding tightly to John’s shirt.

“What are you talking about? I don’t understand anything. Bill please, let me get help first!” John pleaded, trying to pry his hand from his shirt.

“Get me up, I will be fine… I promise, I am fine.” Bill pleaded in return, not letting go.

John had to make the hardest decision of his life. He was just as distraught, and he clearly didn’t understand what was going on. But the expression on his friend’s face, finally made him choose one. “Fine…” He grudgingly accepted, helping him to stand. “Can you tell me what’s going on?” He asked as soon as Bill was sitting.

“The report says the same… the signs are the same … people are vanishing, except this time, it is happening all over the world.”

“Who is vanishing?” “The news just said fifty thousand people worldwide.”

“What news are you talking about Bill? The fucking TV is off.” John yelled, pointing towards the direction the TV hung.

“What?” Bill asked, looking at the side. And just like John said, the TV was now off; a blank screen that held no answer. “The power must have gone out again.” He said, holding on to his temples. As he looked up, he saw the expression on his friend’s face, and he knew he didn’t believe him. I swear… I heard it, the news. There was a directive from the president, there is an app pandemic out there with everyone vanishing without no explanation. The app… it is all real. I didn’t want to say anything. I thought it was all over. But I was so stupid.” Bill explained.

John wanted to believe all he was hearing; he wanted to believe so badly. But – nothing made sense. As he stared at his best friend, he saw a person that was completely different from the one he knew. “What the hell happened to you?” He whispered.

CHAPTER SEVENTEEN

How do you explain to someone something that doesn't exist? Something that seems like a total lie, and yet it couldn't be truer than any other truths you've ever told. The howling wind of misery blew in from the north, sweeping across the city, and sending chills down the spine of those who looked it in the eye. At sixteen years old, most teens are said to be at the peak of their teenage years; times when you either go wild, or you are as free as a bird. For Bill, it was neither.

On his young shoulders was the weight of the world; he was slowly losing touch with reality. He could clearly remember the day he heard the news report; his body remember the fear it felt, his mind remembered how paralyzed it was. But, was it really all a dream?

"Nice kicks." Someone commented, causing him to raise his head.

"Huh?"

"I love your shoes." The person repeated, pointing at them.

"Ah... thanks." Bill replied, forcing a smile. He really wasn't in the best of mood, and small talks weren't his best forte.

"Keys, we need to go!" Another man from the distance yelled, drawing the commenter's attention.

"Geez, I am coming…" He grumbled. "Okay bye." The man said, and just like he arrived, he was gone in a flash. Bill stood in the same spot, watching the man get into a car with the one that called him. The world could be ending, and someone just commented on his shoes. The irony of it made him crack a smile, but the misery that followed made it disappear as soon as it appeared. He wondered who the man was, and what kind of life he led up until that moment. *Would he also disappear one day?* He reflected.

Breathing in deeply, he turned around and walked towards the same direction he had been going; but this time, a new determination was on his mind. He needed to see Marcus; his father's friend.

It took the same amount of time to get to Marcus house. It had the same driveway, same porch; same building… nothing was different. And although he had made the decision on a spur, something was now dawning on him. The man didn't know him. How would he introduce himself? What would he say? Rampant thoughts ran through his head, making him get more nervous. "Auw shit!" He cussed, ringing the doorbell.

Stepping back, Bill shoved his both hands into his pocket; feeling his palms get clammy. It didn't take long for the door to open, and the familiar face of the man who was now a stranger appeared by the door way.

"Hello… Mr. Marcus." Bill greeted, wincing inwardly at his harsh voice.

"Yes… Is there something I can help you with?" The man asked, a crease between his brows. In his eyes, there was nothing but curiosity; curiosity towards a stranger that had suddenly appeared at his doorstep.

"Yes… uhm, can I come in?"

There was an impregnable silence following his words, and although it felt like an eternity, it was only a few seconds.

"Sure." Marcus finally agreed, opening the door wider. He led Bill into his office; the same space that led to a secret room just within reach. "So… what can I do for you?" He repeated again.

"If I told you I have the ability to create a portal that leads into the future and the past; would you think I am crazy?"

"Well yes." Marcus replied without missing a beat, but then added. "But it sounds interesting; can I hear more of that?"

Although Marcus had not heard the full story, neither did he say he believed him, Bill felt true hope for the first time in a week. Breathing in deeply to calm his nerves, he began to tell a tale the man had forgotten. "My friend and I created a mobile app; a portal that could answer our dreams. But little did we know the chaos it would wreak on the world; the implication of messing with time." He started.

Bill told him everything; not holding back any detail. He told him of the algorithms he had used, he told of the fight they had fought together; the answer they had both gotten – everything. Marcus sat in silence, a contemplative expression on his face throughout the entire story. He didn't ask any question, neither did he show any slight. "So you're Jordan's son?" he asked.

"Yes."

"Hmm… you do look like him a bit. Does your father know you're here?"

"No… He doesn't. He doesn't know I am here." Bill responded, swallowing hard the spit that had gathered in his mouth. "He also doesn't remember."

Another moment of silence followed, and as Bill looked up, he noticed Marcus was studying him for a bit. "I know you didn't believe me…"

"No – No, it's not that." Marcus replied, cutting his words short. "I am just wondering if you're really sixteen. I mean… this is, it seems plausible; the algorithms, the modulations… the interfaces trio X twenty four, everything. I am seriously blown away. I need to make a call." Marcus said, standing up from his seat.

"Does that mean you believe me?"

"I don't know…" Marcus said truthfully, shrugging. "But if it is really true; I don't want to regret not doing anything." He said, the corner of his lips tilting into a smile. Then he was gone.

All alone, the silence in the room echoed; amplifying the thoughts in his head that couldn't connect. "What happens next…?" Bill mumbled, resting his head back on the headrest.

CHAPTER EIGHTEEN

As soon as the doorbell rang, Marcus was up on his feet, Bill following hotly behind him. Flinging the door open, they were both met with a middle aged man; much younger than Bill expected him to be.

"You are here." Marcus stated.

"Yeah, yeah." The man replied, walking inside. Bill stood watching from the side, wondering what role the new man would play.

"So what was the emergency? The man asked, folding

"Billy meet Victor, my younger brother; and Victor meet Billy."

"What's going on?" Victor asked, his eyes darting from Bill to Marcus. "You said it was an emergency." He accused.

"It is." Marcus assured. "You might want to sit down for this."

"This better be good Marcus. I left everything to come here, you don't know how much pressure I am under these days."

"I know... I know." Marcus sighed, leading the way into his study. In a different situation, Bill knew he would find their interaction amusing; since both brothers looked nothing alike.

"Different dads." Victor said, almost as if he had heard Bill's thought. Bill couldn't help but raise a brow up. *So, one brother had good hearing, and the other had clairvoyance?* He thought.

Victor nodded his head, a telltale sign for Marcus to carry on.

"What do you need?" Marcus asked.

Bill decided he would rather show them instead of launching into a long tale. "An untraceable laptop." He replied.

"Okay." Marcus replied, heading to the other side. Just like in the past, he began to feel out the wall until he found what he was looking for. The solid wall gave way to the hidden space behind it.

"You've been to my hidden space before, right?" Marcus asked, turning towards Bill.

"Hmm..." Bill responded, looking around. Although he had been here before, the place was still impressive. As he turned towards Victor, he noticed the man's expression remained the same. '*Perhaps he has been here multiple times?*' He wondered.

Soon enough, he was provided with the tools he needed, and he found himself seating in the same place he sat months ago. An undisturbed silence followed, and it was like watching a repeat of what happened the first time they met, except there was now another enthusiast next to them.

"My God... although you told me about it; it is very different seeing it in person." Marcus whistled appreciatively, his eyes following every step John took. "How did you sequence the algorithm to be 6'4 interface of the eighty sixth formats?"

"Code CCPP is the extent of the of the RR1 code, we could bypass the Ender-one file disturbance to another directory information pathway." Bill said, his brows drawing close as he stared at the code lines on the screen. "The NET libraries make a compelling framework for anyone wanting to write cross-platform tools quickly and easily. Running the code from Listing 1-1 should decode both the string "hexagonal octal" Bill responded, repeating the words his brain had memorized. It was almost chilling how even his answers and the questions were the same. It was like witnessing a real loop in time, and Bill could feel his back sting from what he had really created.

"What of the CCPD error... you didn't encounter that while you were writing this?" Victor asked, joining in on the fun.

"The algorithm Artemis helped to bring out from the information bank. Zeus data risk embraced the open source software in ways many people would have thought unimaginable; and with the help of AlPine60 and Beta70.5, we were able to cancel out the CCPD error."

"Oh my God..." Marcus whispered, followed by a long line of cusses as he straightened his frame. "This is just... didn't you loop the recon to the Sqlmap?"

"The Test Post Request with Sqlmap method accepts four influences. The first influence is the URL that will be sent to sqlmap. The second influence is the data that will be in the post body of the HTTP request—be it POST parameters or SOAP XML. The third influence is the value that will be passed in the SOAP Action header in the HTTP request. The last influence is the unique value that is vulnerable. It will be replaced with an asterisk in the data from the second influence before being sent to sqlmap to fuzz. After we bypass the message to the screen we might be able to get a contemporary URL; and once we tested it out, it worked. So we created a Sqlmap Session and Sqlmap Manager objects." Bill answered.

As soon as those words came out, Bill felt his mind go blank, and a serenity wash over his body. But he knew it was all false; it was a state his body had gone into to protect his mind. It was literally the same thing. He was repeating history word for word.

"Just one question... This is a spectacular and awe-inspiring piece of art. What is the problem?" Victor asked, a crease between his brows. "You should be extremely proud of yourself, young man. But who is we? You keep mentioning someone else"

"It's my best friend John..." Bill trailed off, looking at Marcus for help.

"You need to sit down." Marcus said.

"There is more?" Victor asked, raising both of his brows. "Fine, I will sit." He said, giving in. as soon as he sat down, Marcus told him the truth, not skipping a beat. "It's the app that is making people vanish."

"Oh My God!" Victor yelled, standing up almost immediately. "Wait... hold on." He laughed tersely, holding on to his temples. "Let me get this straight, the vanishing cases; the current case we are working on… you're telling me the whole cause is right in front of me?'

"Yes, it is the same one." Marcus said gravely.

"It all makes sense now…" He said in a low voice, his eyes dazed as he thought about something. "They were all kids… the victims, they were all kids." He said, looking straight at his brother. Immediately Victor's demeanor changed, and his face contorted into a ball of rage. "I will under no circumstances be a part of this crime. You should be ashamed of yourself Marcus. Even if the boy knows no better, how come you're still a part of it?" He reproached.

"For God sake! Just Fucking sit." Marcus cussed, pushing his brother back to the chair. "We already deleted it before, right?" He asked, turning towards Bill.

"Yes…" Bill said in a low voice, unable to meet Victor's glare. "I deleted the app with my own hand."

"Then what happened?" Victor asked.

"That's what we are trying to figure out." Marcus replied, rubbing his right hand down his face.

"How hard is it to delete one app?" Victor asked, this time, turning towards Bill.

With slouched shoulders, Bill retold the tale of how hard of a fight they had against time.

"You used LEO? You contacted Sam, from CASSIOPE?" Without waiting for Marcus to reply, he continued. "That cynical bastard couldn't have helped for free. What did you bargain with?"

"I don't remember." Marcus said with a shrug. There was no way he could remember something that he didn't even know took place.

"But are you aware that the satellite used could have the ability to store data received while passing over one part of earth, and transmitting it later while passing to another part of earth, since it is a CASCADE system."

"Since it is a CASCADE system." Marcus said, saying it at the same time his brother. "You don't think Sam used it for..." he trailed off, running a hand through his hair.

"That bastard." Victor cussed, laughing cynically. "I won't be surprised if he did. His soul is so evil that he would definitely remember it all. That might have been the leak."

"It is all speculations." Marcus defended.

"Yeah right." Victor snorted.

Bill looked from side to side, not sure what was going on at that point. "Excuse me...?"

As soon as his voice reached towards them, Victor's phone began to ring. "I need to take this, excuse me." He said, leaving the room in a jiffy.

"What is going on?" Bill asked, turning towards Marcus.

“I think my past self-made a very big mistake.” Marcus groaned, burying his face with both hands. “The friend I called for the satellite link? He might have copied the data from the app when it stored temporarily on the LEO.” He confessed.

“Someone other than me retained their memory?” Bill asked, more shocked than surprised. “And… I… Is that possible?”

“I don’t know for sure, but it is possible… it could be possible.” He replied. “Have you never questioned why you still have your memory, and how come the others don’t?” Marcus asked, finally raising his head up.

“Maybe because I was directly involved with the app?”

“Then your friend and I would have had our memories intact. But it got cleaned off just like the rest of the world. Perhaps time choose its players… who knows; at this point it all sounds like something you would see off a movie or a book. Except it’s not.”

“Except it’s not…” Bill repeated.

“I have good news and bad news.” Victor said, appearing once more by the doorway; breaking the moment. “The good news is, Sam was indeed the leak.”

“The bad?” Marcus asked, straightening his spine.

“Interpol got involved. They might be here in any minute. Sam told them he got it from you.”

"Oh My God... Oh My God..." Marcus groaned, burying his face with his hands once more.

"There's more... its worse." Victor added, his voice somber. "Grace just told me that Key called her, and the new reports says one hundred and eight thousand, five hundred people vanished the same way around the world. The exact same signs; no other clues left behind."

The harsh sting of death washed over Bill, and it felt like he was falling through the crack of time; an endless void he would never climb out of. Just when you think you are at your worst, Life proves it otherwise. He would never have fathom, not even in his wildest dreams that this could happen. Although earth's population was in the billions, one hundred and eight thousand was still a big number for his teenage mind to process. He was after all, still a young kid, and a good one that meant no harm.

"It's beginning to become an app pandemic." He whispered through his broken state of mind, and almost immediately, the doorbell rang.

Jordan closed the door behind him, his footfalls echoing as he walked down the corridor. "Bill, I am home." He called out to the open space, expecting an answer from his son; but his voice never came back. "Bill?" He called out once more, wondering if he was at John's. 'That kid is never home these days.' He grumbled. He removed his phone from his back pocket, and just as he was about to speed dial Bill's number, another call came through. The number belonged to an old friend, and a mix of surprise and shock were etched on his face. “Marcus? What… It is been so long. How have you been?” He greeted with gusto.

“Yeah... It is me.” Marcus replied, his voice an opposite of Jordan’s enthusiasm.

“I hope nothing is wrong…?”

“Code orange.”

As soon as Jordan heard those words, his demeanor changed, and the lighthearted expression on his face turned grave. “Where should I meet you?”

“I will send you an address… and…” Marcus trailed off, hesitation in his voice. “You need to be prepared… it has something to do with your son.”

CHAPTER NINETEEN

Jonathan Jones, can you please report to the principal's office. Now." An announcement was made over the school's speakers, drawing the attention of every student present.

"John, John… Isn't that you?" A friend called out, pointing in the direction of one of the speakers.

"Me…?" John repeated, frowning. "Why?"

"How would I know?" His friend shrugged. "You better go."

John was baffled, walking towards the direction of the principal's office. He had not done anything to warrant being called, so what was going on?" He wondered. Bringing out his phone, he checked the time, biting his lip as he contemplated something. Bill has been missing from school for three days. John had tried calling his phone; but it was switched off. He had also gone to his house; but neither Bill nor his dad was around. He was getting more worried, since nothing like this had ever happened. Not once have they ever upped, and left. But then again, no news was good news.

Counting three in his mind, John knocked on the principal's office.

"Come in." A voice responded from inside.

Nothing seemed different, except the secretary wasn't around. As John stepped further in, he called out to the Principal. "Mr. Wilkins… It's me… John?"

"Here" A man replied, opening the door to the main office.

John paused, staring at the man. He wasn't a familiar face, and he definitely wasn't the principal. Mr. Wilkins was a sixty something year old man, who always dressed in checkered shirts that had short sleeves, and a tailored suit that was grey, brown, or black. The man in front of him was neither of the qualities he just thought about. He looked to be in his thirties, and he was currently wearing a black suit. "Oh, I am looking for the Principal." John stated in a matter of fact manner. "Is he around?"

"I know." The man replied, opening the door wider. "Come in." The man responded, not directly answering his questions. Tentatively, John entered the room, and discovered there were two of them.

"Sit down." The first man said, before proceeding to sit down at the other side of the table; alongside his partner.

"Yeah…" John said in a doubtful tone, sitting down gingerly. "Uhm… so… Is there something I am supposed to be doing?"

"Well… My partner and I are from the Interpol."

"Interpol?" John repeated, shocked. "The Interpol we all know?"

Instead of answering, the first agent brought out an ID card, sliding it across the table to John. Picking it up, John stared at the picture on it, then back at the man who gave it; before pro9ceeding to read the other information. "Seems legit…" He mumbled, returning the identity card. "So you are like… I don't know… secret agents or something?"

"Contrary to frequent portrays in popular culture, Interpol is not a supranational law enforcement agency. We are an international organization that functions as a network of law enforcement agencies from different countries. My partner is from France, and I am from the country right here." The first man replied. John wondered if this was all a prank, but he decided he would rather waste time here than go to class. He was also going to label them as agent one and two since they both didn't offer their names. "Okay… cool."

"We will know proceed the questioning."

"The questioning?"

"Are you going to repeat everything we say?"

"Ah ha… sorry about that." John replied apologetically, scratching his nape. "Yeah… sure, whatever you want to know." He shrugged.

"Do you remember anything that happened two months ago?"

"Two months ago… Like what?"

"Like the time you fall sick… or had a nightmare… or the time you vanished?"

“What?” John laughed tersely; wondering if Interpol agents could also crack jokes. “I don’t even remember what happened yesterday.” He joked back. “But like seriously, I never get sick. Like ever. I don’t even catch a cold.” He scoffed.

“You also don’t remember vanishing?”

“Uh…” John muttered, raising his left brow. “Vanishing like how? Are you talking about vanishing… like, the word?” He said, hoping for clarity.

“Yes, like the word, Mr. Jones.” The Agent repeated, enunciating every word.

“Is that even a question… like?” John shrugged, breaking into a short laugh once more. “Won’t I have to die for that to happen?”

“Technically, but you didn’t.”

John broke into a full laugh. He didn’t know who these men were, but they were killing him. “Okay, that – that was a good one.” he said between laughs; but one look at their serious face, and his laughter died out. “Ha… okay, I will shut up now.” He said, making the zip sign with his mouth. But little did he know that it was only the beginning of a long line of questions. John was beginning to feel weird, and the more questions he was asked; the more confused he was. “Can I go now?” he asked, pointing towards the door with his left thumb. “I would love to chat more, but I have three more classes today.” He said. For the first time in his life, John was willing to take a boring class to the interrogation he was currently going through.

"Classes are the least of your concerns right now, Mr. Jones." His partner replied, speaking for the first time since John entered the room.

"What?" John responded.

"You heard me right, Mr. Jones." The man replied.

John was beginning to get irked out by the way they kept calling his last name. "Look, what do you want? I don't even know why I am here. I just... I don't know." He replied, looking down at his clasped hands. "You can't detain me here. I have basic human rights." He mumbled.

"Yes, you do have them. But I wonder if we can say the same for a criminal."

"Criminal, you have to be shitting me?" John cussed, reaching down to grab his bag." I am leaving." He said, standing up.

"Mr. Jones you shall sit back down." The first agent said, his voice containing a veiled threat.

"No, I don't want to." John yelled. "All you did since I came here is ask me questions I don't want to answer. I don't even want to be here." He said, yanking his backpack against his arm. "I didn't even do anything!" He protested, his voice sounding shrill. By now, the panic was now apparent in his voice, and his heart was beating louder; slamming against his ribcage.

"We will soon be done. Please seat back down." The first agent said his voice gentler than before. An impregnable silence ensued, and it almost seemed like John would rebel. John wish he could storm off, but one look at their faces, and he knew these men were not joking; they made sure to show it.

"Fuck!" He cussed, wiping the corner of his left eye. Grudgingly he went back to seat.

"Have you heard of a mobile app called the portal?"

Their question was met with silence. John decided he wasn't going to reply to anything they said.

"Do you know the app has been downloaded one point five million times?" It was the same this time, John wasn't answering.

"Do you know Billy Martins?" He asked.

Hearing Bill's name, John looked up. "Yeah..." John said hesitantly, his eyes darting from one man to the other. "What about him?"

"Who is he to you?" The second agent queried.

"He is my buddy... Uhm... he is my best friend." John answered.

"Are you aware that Billy Martins?

“Are you joking right now?” He laughed, shaking his head. “That’s not possible.”

“Why do you think we are here?” The second agent asked, shrugging.

“I don’t know. But you have the wrong Bill. Not the Bill I know.”

“That young man created something that is currently on the verge of causing a worldwide epidemic.”

“What are you talking about? Why do you keep saying things that are not true?”

“And because of what he created; a world summit would be held next week to decide the fate of the world.” The agent continued, ignoring John’s words.

“What the hell are you saying?” John repeated, banging his hand on the table. He was frustrated that they were not listening to him, and yet – kept spitting facts that were utter bullshit.

“Are you really sure you don’t know anything?”

“Of course I am very sure! Bill would never…” John trailed off, his eyes dazed as his mind travelled back to a scene he had completely forgotten. He remembered the moment Bill had looked like he was having a panic attack; and the words he had said.

The whole world is vanishing and I don’t know how to stop it.” Bill cried, holding tightly to John’s shirt. “Get me up, I will be fine…

I promise, I am fine." Bill pleaded in return, not letting go. "The report says the same… the signs are the same … people are vanishing, except this time, it is happening all over the world."

"Who is vanishing?"

"The news just said fifty thousand people worldwide."

"Mr. Jones? Are you listening?" The first agent asked, tilting his head to the left as he stared at John.

"Yeah… yes... Yes." John repeated, blinking rapidly as he tried to clear his thoughts. "I don't know…" He said, unable to meet their gazes this time.

"You will have to come with us." The agent added, arranging the papers on the table between them.

John wanted to ask if he could refuse; but he could almost hear their answers already. "I need to call my mom."

"We already did."

"What of Bill?"

"He is also with us."

Bill wasn't sure how long he had been kept here; he wasn't sure of the day of the week, neither was he sure where exactly he was. When Interpol had arrived at the doorsteps of Marcus' house, Bill had thought they would treat him like a criminal; but it was quite the opposite. Apart from not being allowed to see anyone, they had kept him in this building since his arrival; providing him with everything he needed. They had also taken his phone, any tech stuff he could lay his hands on.

Rolling on to his side, Bill wondered how his dad was, if he was aware of what was going on; and also…John. "It wasn't supposed to end this way…" He muttered, his throat burning from the emotions that clogged his throat. The sound of someone knocking broke into his thoughts; clearing the fog that surrounded his head. Bill wondered what would happen if he simply ignored it, but whoever was at the door was very persistent.

Grudgingly giving in, Bill stood up, flinging the door open. "Hello." The person in front of the door greeted, causing the creases in Bill's brows to increase.

"You are…" he trailed off, hoping the man would fill it in.

"Detective Key." The man replied, pulling his hands out of his pocket. "Wait a minute… aren't you the kid with the nice kicks? At uhm… PACO? Small world uh…?" He stated, tilting his head a little to the side.

"What do you want?" Bill asked.

"You have no idea how much of my connections I had to bring out, just to be here."

"Why?" Bill queried, a bit perplexed that a random stranger was trying hard to meet him.

"I will buy you lunch at the cafeteria downstairs, and then I would tell you why."

"I am not hungry."

"I remember the moment in time where everyone's memories vanished."

"What?" Bill asked, not sure he had heard correctly.

"You heard right." Key added.

Gripping tightly to the door, Bill was unaware of his knuckles turning white. 'How many of those people were out there in the world.'

"Do you want to come downstairs now? I will tell you all about it." Key asked, pointing towards the direction he just came from. Bill wanted to refuse, but curiosity was a strong thing. He wanted to know how their experiences differ. He wanted to know if they were connected ion one way or the other, or if they had both done the same thing. So, Bill followed Key.

But he never had the chance to ask him those questions he truly wanted to know. As soon as he got downstairs, Bill came in contact with someone he had never expected to see so soon; maybe never considering all things.

"John?! He yelled, his voice bringing everyone's attention to him; including Johns'

"Bill?" John responded not believing his eyes. He wondered how someone's thoughts could materialize into the real thing. And almost like a scene from a movie, Bill broke away from the man that held his attention a moment ago, and into the company of the one that was a brother to him. "It really is you…" He said, pushing his shoulder. "You are really here… how?"

"Interpol came to school."

"John I can –"

I heard the whole thing…" John said, his voice lower than before. "Marcus came to see me. He told me everything you told him. But, for how long?"

"John…" As soon as he said his name. Bill watched as his friend's face turned from ecstatic to emotionless.

"Answer me! For how long were you going to keep me in the dark?" John shouted, his voice echoing and bouncing against the wall.

"I was trying to protect you." Bill replied, shoving his hands into his pocket. He knew is he didn't, he would not be able to hide his shaking hands.

"Protect me?" John scoffed walking closer to where he stood. "You wanted to protect me? And this was the way to go about it?" He said in disbelief, shaking his head.

"John, please just listen to me…" Bill said pleadingly.

“Listen? Why should I listen to someone that can look you in the eyes, and keep lying to you?” John yelled, pushing Bill’s chest with enough force to make him take two steps backward.

“It all happened so fast, I didn’t know what to do, or how to act. And I was scared. I was scared that if you remember, something could happen; like a glitch in time. ” Bill responded, trying to explain his feelings the best he could.

“It happened two months ago Bill. Two months! And all those times, there was not even a single chance for you to tell me what happened? You are best friend! You can’t keep things like this from me!” John yelled louder, pushing Bill with more force.

Bill stumbled backwards once more, hitting his back against the edge of the table. At that moment, his mind was a myriad of emotions. But the strongest of them all was Anger; a pulsating emotion that burst through with a ball of energy.

“I once tried…” He muttered, straightening his frame.

“You once tried? When? How?” John accused, throwing his hands in the air. “Don’t even try to bullshit your way through this.” John added, his voice showing anger and pain as he stared at the one that had caused it. Bill was not only his best friend; but a brother he never had. He was family.

He remembered the moment his heart had almost stop beating when he heard he had helped create something that could destroy the world. He was mad at Bill for not sharing his pain, or reaching out to him when he needed to the most. For bearing the whole burden, and thinking he didn’t want to protect him too.

"I tried telling you; you were the one that didn't believe me!" Bill yelled, pushing back at John's chest. "I tried…" He repeated, but this time in a lower voice.

Meanwhile…

Jordan could barely suppress shaking his leg in impatience. He was waiting… waiting for the agent that was assigned to them. The man was twenty minutes late; and each passing minute was like swallowing nails. Although he didn't show much, since Maria was right next to him; and at the moment, he was her own pillar of support. But inside, it was a different thing for Jordan.

His emotions were in chaos.

He wanted to yell, he wanted to break things, and demand for his son; but neither of that would bring him back. It had been ten days since the last time he saw his son; for Maria it was the same thing.

Ten minutes later, the door finally opens, and the assigned agent comes through. All the fight Jordan was building up disappeared like mist, and at that moment all he wanted was to know how his son was. By now, all three were standing.

"Mr. Martins, Mrs. Jones. Please sit down." He said, pointing to the chair they had been occupying earlier. Following suit, he began. "I am Agent Maine from HQ office, France." He introduced.

"Nice to meet you."

“Nice to meet you.” Both parents chorused, hoping the discussion would move along. Jordan eyed the thick binder between them, wondering if they were going to go through all of it. But instead, the agent followed through with questions.

“Were you aware that your sons were making something that dangerous? This is something that would have taken years, and neither of you noticed?”

“Hmm… Your sons at the moment are more or less international criminals; but at the same, they are our treasured guests. We need their help to solve the disappearance of those that vanished.”

At his words, Maria burst into tears.

“I am sorry to tell you this, but your sons would have to come with us. They will have to be registered into the program”

“Why… what’s going on?” Maria asked between sobs, tightening her grip on Jordan’s hand as she waited for new information.

“They have been registered as targets on the black market.”

“What do you mean they have been registered?” Jordan asked, finally saying something.

“To put it in simpler terms; they have a bounty attached to their names.”

CHAPTER TWENTY

John shook his head as he stared at his best friend; his mind trying to deny the fact that Bill might be right. His brain knew it all, but his heart wasn't ready to accept that. "I guess you would rather trust a stranger than trust me." He said, saying selfish words he didn't mean. John winced inwardly as he saw guilt darken Bill's face; but it was too late. He couldn't take back the words he already said. All that was left was his ego and pride that were barely standing. The human heart was a very strange thing. It was clear that John cared about Bill, and yet – the instinct to hurt the person who hurt us first, sometimes cloud our judgment, and the punishment we dish out is always double of what was given onto us.

"Mr. Jones, we have to go" a voice from the sidelines broke the moment, drawing John's attention towards his left; it was the federal agent assigned for his safety. John felt a bit relieved that they had to leave. He didn't know what to say, neither did he know what to do next; so taking an exit was highly welcomed. Nodding his head, he turned around, about to walk off into the opposite direction. He could feel his back burn from the gaze that drilled a hole in his back; but he didn't turn back. Instead, he walked off with rigid shoulders, his face twisted with regret and his heart mourning the words he should have never said. He felt like the biggest jerk on the planet at that moment.

Bill watched the back of his best friend disappear into his distance, and although he urged his legs to move, they were deeply rooted to the ground. "Are you not going to go after him?" A voice said, breaking into his thoughts. Looking to the right, he became aware once again of the presence that was by his side all along. "Sorry." He muttered, running a hand through his hair. "I think you need to come another day. I don't feel like eating anymore."

"I wish I could say okay, and leave. But – If I go now, I might never see you again."

"I will still be here anyways." Bill argued, dropping his hand back to his side.

"Not for..." Key trailed off, pausing the words he was about to say. "Just give me thirty minutes. We don't have to eat." He added, shrugging.

Bill knew he would like nothing more but to disappear into the room he had been assigned to. But once again, his curiosity reared its head. The man in front of him was the first person he had met, that had memories of a time no one else could remember. The third person was under CIS custody, and the likeliness of them ever meeting was very low. "Alright." He sighed, pointing towards the empty chairs next to a food court. "Let's sit over there." He suggested, heading towards the direction without skipping a beat. It didn't take long for them to settle down, and soon enough, they were both sitting opposite each other.

"When did you realize something was wrong?" Bill asked, asking a question he was most curious about.

"Three days after the incident." Key replied, lightly scratching his chin. "Same routine, same people, same places... everything seemed normal, and that was the scariest part of it all. I was living in a world that was normal, and yet nothing seemed normal." He explained, a crease appearing between his brows.

Bill felt apologetic almost immediately, causing his shoulders to shrink.

"The day it really hit was the day the last victim disappeared. Time was repeating itself, and I was the only one that could notice. I had a moment of vertigo as I stood in the middle of the road. I was staring at familiar things that all seemed strange. At first, I thought I was crazy. Lost my mind... lost my senses. But it turned out I was not crazy. I just happen to remember a past that ceased to exist." He said, drumming the fingertips of his right hand on the edge of the table. "Not going to lie, it still sounds too crazy to believe. "Victor. He was the first real truth, and I grabbed that line; and here I am." He replied, shrugging.

"Are you here to blame me?"

"Blame?" Key repeated, raising his head up. "Blaming you is the last thing on my mind. I am here because of something else." Key said, looking directly at Bill. "I need to know what really happened. You need to start from the very beginning,"

Bill stared at him, the corners of his eyes twitching for a split second. "I guess it is time for the interrogations to begin. I knew this was coming... and here we are." Bill muttered.

"The more you tell me, the more I can help you."

At his words, Bill looked up once more, frantically looking at the expression of the man that was in front of him. "What of my dad? Why is John also here? Unlike us, he doesn't remember anything. He is just as innocent."

"Tell me all you know, and I will tell you all I know. After all, I need your help to find the missing people." He said.

At his words, Bill looked down. How was he going to tell someone that was looking at him earnestly, that he couldn't help? He had no idea what was going on; with the app, the time glitch, his life, his dad, john... the missing teens; he had absolutely no clue on how to help them. He had once tried his best, and yet – the end result had everything falling apart "It is almost as if life is mocking me..." He mumbled, resting his arms on his table. "There is nothing I can do." He muttered, his eyes darting from side to side, unable to meet Key's gaze.

"I know you once saved them."

"You are right, I once did. But then it happened again, and this time – even worse; more people disappeared. It was like time was telling me not to mess with it..."

CHAPTER TWENTY ONE

Present Time: 8:45am

Location: National Intelligence Service; NIS

Someone once quoted "The course of life is so unpredictable, that no one can write his autobiography in advance. It was a place filled with ups and downs. And the only thing we know about the future is that it will be faster changing, and more unpredictable than the present. Another thing that made life unpredictable was the capricious nature of humans. Humans are inscrutable; infinitely unpredictable. This is what makes them dangerous. Their actions fill the blank pages called life, staining it with the blemish of their sins.

Nar Lee Octavo checked his watch for the thousandth time, cussing beneath his breath as he hurried towards the building that stood tall, few meters away from him. He cussed the person that thought it was a good idea to fix a meeting so early. Nar was in no way a morning person, and today was no different. He wished he could get coffee; but the queue at his favorite spot was no joke, and he couldn't afford to be later. As he moved along, his eyes scanned the circumferences of the building, noticing the amount of security that had been put in place for today's meeting. Today's meeting was a summit with top officials from all over the world to deliberate on an emergency matter that was slated the highest emergency.

Nar could finally breathe easily as he took an empty seat at the very back. For some reason, he was glad the summit had not begun. He would hate to walk in the very moment the meeting had started, only for everyone to turn their attention towards the new person. Adjusting his tie, he looked around and noted a few important figures. He could spot the ambassador from India, Germany and Netherlands. As he looked towards the opposite direction, his eyes zeroed in on the stage in the middle of the conference hall. Almost on cue, two men dressed in suit walked towards the stage, signaling the beginning of the summit. Leaning forward, Nar drummed his fingers on the small table in front of him, and knew this could end in two ways. It could either be good and short, or be a long line of boring talks. But nothing in his thirty years of service could prepare him for what was to happen next.

"Good Day everyone, I am Xavier McCullough; the Chief of the Intelligence service, and the host for the emergency summit 2022. This summit unlike the previous years would probably be the most important, and the most life changing discussion we have ever held; and that's why it was coded the highest level of urgency. It is an international disaster response. I apologize for the short notice, but there was no choice."

At his choice of words, everyone in the hall stood at alert, wondering and pondering what exactly was happening. Nar was no different, and he could hear the blood rushing through his veins as the entire room stood silent. But whatever he had learned throughout his fifty years didn't prepare him for what was to come next.

"About two months ago, a string of odd disappearances began in the United States, and gradually it spread across the country; and finally outside the country. At first, our investigations led us to believe it was random kidnappings with no connections; but as the signs left behind repeated itself – it led us to a much bigger canvas, and the investigation took another turn. We thought we were dealing with a serial kidnapper; or an organized mob, that dealt with cyber crimes and kidnapping. But we couldn't be more wrong." The Chief stated, causing murmurs to rise amidst the crowd. "The next words I am about to say might sound like something I picked off a sci-fi novel, but this is the truth, and the only truth we have." He added, scanning the crowd, and holding every single person at attention."

"But do we have a culprit now?" The Indonesia representative asked, drawing everyone's attention towards him.

"I will be getting to that, but first I need to explain what we are dealing with. We are currently dealing with an invisible enemy created by the glitch in time. And by glitch in time, I mean instability to the inevitable progression into the future with the passing of present and past events. A mobile app was created to access the past and the future, to change an associated part of the second law of thermodynamics; the physical and psychological result of increasing entropy. The reason for those disappearances is due to an invented app that makes them stuck in the loop of time." The Chief explained. At his words, another round of hushed whispers rouse amidst the crowds, leaving behind many questions that were yet to be answered.

Nar was no different from everyone in the room. Everything he had heard so far felt like something from a sci-fi movie; just like the Chief had stated earlier. It was almost comical to the whole situation, and the level of seriousness was barely sinking in. It wasn't the first time something strange had come up on discussion, but it was the first time something ridiculous was brought forth, and called an emergency. But there was just something that niggled at the back of his mind; something that left him with an unsettled feeling

"As I said earlier, mobile app was created to access the past and the future, to change an associated part of the second law of thermodynamics; the physical and psychological result of increasing entropy. The app is something like a portal. A time machine that could give us the benefit of changing history, and eradicating the regrets we once have. But that's only the good part, the bad part is this portal is currently not under control. It is self-reliant, and is operating under a set of codes we are currently unable to access. We are trying our-"

"Are you saying that there will be more disappearances?"

"What Country would be most affected?"

"America started it all, what have you been doing to stop this?"

"You have not told us who the culprits are?"

"Is there something you are not telling us?"

Multiple questions broke out from the crowd, cutting short the Chief's explanation.

Nar felt a bitter taste fill his mouth at the self-centered questions that was being asked. Instead of thinking of possible solutions, they were more worried about the country would be hit first. Men were nothing but selfish creatures that only thought about themselves; he knew this fact, but they always managed to surprise him every single time

“But is there a way or a means we can stop this?” The official from Japan asked, his voice booming louder than every other person. His question echoed the fears that were deeply rooted in everyone’s heart, causing a tension so thick that a knife could cut through to fill the room. And at that moment, the pause that followed felt like an eternity; but in actuality only a few seconds.

“About that, we currently don’t have a way to stop it. And this could mean, the whole world could disappear at any moment.” The NIS Chief replied, causing another loud stir in the air almost immediately.

CHAPTER TWENTY TWO

Bill stared at the agent in front of him; an unreadable expression on his face. For the past few days, he had been having guests; and so his appearance was nothing new. "Is something wrong?" He asked.

"You have a new visitor." The agent stated, and almost immediately, he began to walk away; a sign for Bill to follow. Closing the door behind him, Bill followed silently, their footfalls echoing down the hallway. As soon as they got to the front of the closed door, Bill's heart began to pound. "Who is it?" he asked, trying to stop the nervousness that was slowing permeating his whole body.

But the agent didn't reply, instead he opened the door, and revealed the mystery visitor that made his heart jump. Inside the room was a man hunched over, focused on the table he rested his arms on. As soon as bill saw who the person was, the nervousness he had felt earlier disappeared.

"Dad!" Bill yelled, happiness radiating in his voice as he called out to his father. The man adjacent to him looked up, and within seconds his face mirrored Bill's expression. "Bill" His father cried out, opening his arms out. Bewilderment, happiness and worry ran through his expressions as his son hugged him in return. "I am so happy to see you." he finally said.

“I am so happy to see you too” Bill replied genuinely, his throat clogged as he hugged his father tighter. They stayed that way for a few minutes, until finally they had to part.

Sitting down across from his son, Jordan scanned Bill’s face, worry and concern etched on his own features. “You’ve lost weight.” He commented.

“I missed your cooking.” Bill said jokingly, the corners of his lips tilting up into a smile.

“What do you miss about that?” His father replied, a grin on his lips. “You know I can’t cook.” He added.

Bill chuckled at his father’s words, lowering his head. “You make the best steaks.” He muttered. Bill knew deep down that they both wished to say something else; but the words were too heavy to be said. Their small talk was just the way of prolonging the inevitable.

“I remembered to water the plants.” Jordan said, scratching his nape. “So, everything is fine back home.”

“Dad…” Bill called out.

“I also cleared the garage last weekend. I thought I might as well get to it.” Jordan continued, pretending to hear the word that broke his heart into a million pieces.

“Dad…” Bill called out once more, his voice breaking. “I am sorry.” Bill said remorsefully, unable to meet his father’s eyes. “I am sorry.” He repeated.

Reaching for the space between them, his dad warmly placed his hand on his; drawing Bill's attention towards their hands. "You are a good kid, Bill. No matter what anyone else says, I believe you." His father said softly. At his father words, it was like the dam holding on to his emotions broke down, and the tears he didn't know he was holding began to flow. He didn't care about how embarrassing he looked, he didn't care that there was an officer in the corner, monitoring their every words, he didn't care that there was hidden cameras. He didn't care at all. Holding on to his father's hand, he broke down and became a mumbling mess. Jordan could hear the occasional 'I am sorry.' And each word was like nail being drilled into his heart.

Jordan stayed by his side silently, tears rolling down his cheeks as he bit down on his lower lip. He was frustrated, sad, and his heart was in pain. There was nothing more painful than knowing you couldn't help the people you cared about, and all you had to do was watch as they suffer. Jordan felt he had lost the most important thing and that loss filled him with emptiness like none he had ever felt before. That loss made him cry his heart out, feeling him with agony and hatred for his weak self. He was sorry to his dead wife, who had passed away when Bill was born. He was sorry to son that had a father that couldn't help. Sniffling, Bill looked up, and through blurry sight, he caught on to a sight he had never seen in his life. His father was also crying. It broke the already shattered pieces of his heart, and he wondered what he really did wrong. Why it had to be him, why he had to be the one suffering, making his parent sad.

"You dad must look like a mess." Jordan scoffed his voice apologetic.

"Who told you everything?"

"Marcus. He called me… but by the time I came over, it was too late. He told me you were with them, and they didn't allow me to see you until today."

"Mrs. Jones…" Bill said, almost too scared to ask.

"She knows. She is also here. They came to pick us up this morning."

Looking down, Bill's mind became more unsettled. "They never tell me anything." He started. "Since the time I was brought here… no one has asked me anything. I don't know what is going on dad. I don't know if…" he paused, swallowing hard, the spit that had gathered in his mouth. "I don't know if people are still disappearing." He finally said, his eyes darting from side to side. "What is going on?"

"Everything will be fine."

At his father's words, bill looked up. "Did they tell you anything else?"

"Yes… the same agent that visited us the very first day. He said you and John will both be asked to help decode the app. But first you will be transferred into a safer facility."

"Safer facility?" Bill repeated. "Where?" He asked.

"A facility that can keep you safe." Jordan replied. Shaking his head, Bill held on tighter to his dad's hand. "What about you?"

Jordan bit down on his lower lip, and for the first time since he saw his son. He couldn't say a thing.

THREE DAYS LATER…

The shadows of the night danced happily away; playing hides and seeks with the lights. In the darkened office, a man hunched over his desk, barely moving as he concentrated on whatever he held. On a closer look, it seemed like he was holding on to a picture; the only source of illumination being the lamp on the table.

Nar stared at the picture of the young teen he was to escort the next day. The boy had brown curls, with green eyes that gleamed as he smiled at the person that was behind the camera. Placing the picture aside, he opened the file he had also been given. As he scanned through the file, a deeper crease appeared between his brows. The kid was like every normal teen; had no troubled past, did well in school, neither did he have a criminal record. And it just so happened that he was the same age as his second son. There was nothing on the file that indicated the kid had the mind of a criminal.

Leaning back on his swivel chair, the wheels pushed back a bit, inching him away from the table. Sighing, Nar began to massage his temples. He sincerely prayed everything went well. His cell rang at that moment, sounding louder in the echoing silence, and startling him a bit. Cussing beneath his breathe, he picked the call, not checking who it was.

“Hello.” He said gruffly.

“When are you coming home?” The voice replied, and he realized it was the voice of his wife.

"Oh… Honey. I don't think I will be coming home anytime soon." He groaned. His job took him most times away from home, and fortunately he had been blessed with a patient wife. They had their trying times, but together they managed to pull through. "Perhaps next week?"

"Okay. Remember Henry's birthday is next week. Don't miss it." She said, referring to his second son.

"I won't." He promised.

"Be safe honey."

"I will." He replied, the line going off.

Tilting his head backwards once more, he stared at the ceiling, watching the shadows dance across the room.

CHAPTER TWENTY THREE

The sky above stretched out as far as the eyes could see; blue and unblemished by the sins of the world. By its side was the golden companion, streaming through and casting its glow across the earth. It seemed like a good morning.

"We are traveling separately?" He asked, looking towards the other car. At that moment, his best friend turned towards his direction; and although they had not spoken for a couple of days, it was like all the animosity had melted away, and all that was left behind was the deep bond they shared. He smiled, and rolled his eyes, and his best friend did the same, eliciting a smile. "Can I go over?" He asked, turning back towards Nar.

"I am afraid you can't do that. We need to leave on schedule." Nar responded.

"It won't take long." He pleaded. But all Nar did was shake his head. Sighing, the young teen ran a hand through his hair, his eyes darting back and then back to Nar's face. "Why are we traveling separately, again?"

"For security reasons..." Nar replied, tapping on to the side of his ear where a small piece of tech was attached to. "We have to leave." He said, pointing towards the car they were going to ride.

With a short breath, he strode towards the car and got on it alongside three other people beside Nar; the driver, the agent riding shotgun, while at the back seat – Nar sat by the left, the other agent sat by the right, and he was positioned in the middle. "For security reasons." Nar repeated, and just like that, their journey began. As soon as time was up, and the cars were ready to move, he could feel blood rushing through his ears, and the slowing of his pulse as cold settled over him like the callous tingle of death – but strangely, he was alive and very much aware of the environment.

Fidgeting for the thousandth time, he leaned forward so he could see what was going on outside the car; a sense of unfamiliarity dawning on him. By side of the roads where buildings with flashing signs written in languages he could not read, and faces he had never seen. Slowly, he settled in for a journey towards an unpredictable future.

Thirty minutes into their journey, fate decided to throw a curveball their way. It didn't take long for him to notice something was wrong. As he turned to his side, he noticed Nar was listening to whoever was speaking on the other side of the ear plug, and by his brows, a deep frown was etched. "Copy." Nar muttered, turning towards the driver. "There is construction work on the third exit; we have been directed to move along the fifth axis." He said.

"Copy that." The agent replied, turning the car to the right. Nar knew all their bases had been covered, and yet, he couldn't help but feel nervous. His mind was screaming out a warning sign he couldn't ignore, and the hairs on his arms were prickling. Reaching for the gun tucked in his waist band, he brought it out, drawing the attention of the young teen that was beyond nervous.

"Is something wrong?" The young teen asked, clasping and unclasping his hands as he waited for an answer. Nar pursed his lips, wondering if he

"Nothing is wrong, just a precautionary –" Before Nar could complete his words, the impact of a truck crushing into them sent the car flying. The impact of the crash was considerably greater in the car, because the mass of the truck was much larger. Their car was more or less a baby pitting against an adult. Every single glass window shattered on impact, turning into shrapnel as the occupants of the car were jostled like paper in the bin. After turning three times, the car finally stopped, lying on its hood. Steam doused from the engine, filling the car with smoke, and the stringent smell of acid.

For a minute, every single occupant held no heartbeat, until finally a gasp broke through. Huffing and puffing as his lungs were greedy for air, the young teen continued to gasp. He was still in shock, and his brain was still processing what had happened. He had not braced himself for impact, and so his whole being was in entire shock. Blood was streaming down the side of his face; his hair matted with sweat and blood. Silently sobbing, he unbuckled his seat belt, falling on his head with a thud. Groaning, he reached out for the light beyond the shattered glass. The young teen scrambled out of the overturned car, barely wincing as the glass shards on the road penetrated his arms.

For what felt like an eternity, but in reality few seconds, the young teen snapped back to the present, blinking rapidly to clear the blurry vision away from his eyes causing his ears to ring out in protest. And for a second there he was lost in my own movie; where the main character is at the climax of the story, where the action is about to get down and just stands there, staring, and then everything slows down.

I stood that way for a few minutes, but this was real life, and reality was slowly sinking in.

With shaky legs, he moved towards the other car that lay in a wreck behind. His ears began to ring, and the static screams throw him off balance. It sounded like the whole world was shutting down on him. He screamed as the horrific and colorful sight behind finally registered in his brain. He knew he did because he could feel his mouth open, and his vocal cords vibrate, but he heard nothing. Almost on auto pilot, his legs began to move forward; urging him on to the truth he wasn't ready to bare. It was more like a message from his brain to his body—danger: run away!

But he didn't.

He slowly approached, the trail of his own blood following behind. As he reached the car, he crouched low, and just like he feared the most he saw the result of the attack. In a pool of blood laid his friend, shot and not breathing.

The young teen burst into laughter, a short high pitched laugh that bordered his insanity; the sound of his mind breaking into little pieces. He had never felt more cold, sad and alone in his life; up until that moment.

CHAPTER TWENTY FOUR

There are moments when things don't seem real. It feels like you're out of your mind, and you're watching things from the sidelines; out of your own body. For John – it felt that way. His mind couldn't comprehend why all these had happened in the blink of an eye. His mind couldn't comprehend why his friend was bleeding out, and not breathing. As if to reaffirm he wasn't dreaming, he slowly got on his knees and reached out to touch the body that looked so cold. His hands were trembling, unsteady as they reached for the truth. As soon as his hand latched on to Bill's, his high pitched laughter slowly melted away, and all that was left was nothing. The fog around his heart slowly melted away, and the pain that followed ripped through his heart. "Bill." He said, his voice sounding like he was winded.

A tear, then double, and finally the whole dam were open. He was openly sobbing; the tear dripping down his chin and onto his shirt. His heart knew pain. Almost as if to comfort his crying friend, Bill twitched at that moment, and through his blurry vision, John could see it. Hope was a scary thing. It was a drug that intoxicated the mind, and changed your whole outlook on life. That little twitch instantly changed his demeanor, and his thought process changed into a different direction. Pushing himself up, the first thing he did was check if there was anyone that could help him. "Help!" He yelled, frantically searching for that line he was desperately looking for. He had no phone; he had no other way of calling the emergency line. His phone had been taken from him the moment he arrived in the first facility.

Spotting another driver who was stopping due to the obstruction in front, John took that chance to run to him. "Help! Help!" He yelled,

"Please can you call 911. A lot of people have been badly injured." He pleaded, pointing towards the direction he had come from.

"Yea – yeah… oh my God." The driver whispered, whipping out his phone. John watched impatiently as the driver called the emergency line, explaining to the service provider on the other line. "Yes, the u-turn to the fifth axis… yes… he is…" the driver looked up. "I can only see one survivor." He continued. "Yes…You need to also sit down, you've been badly injured." The driver said with worry, his eyes scanning his blood stained shirt, and the blood that was by the side of his head."

"Soon." John promised, hobbling back to the direction he came from. The first person he checked on was Bill, and fortunately he was still breathing; he could feel a weak pulse by his throat. Pulling his shirt off, John made a bandage out of it; wrapping it around the bleeding part of his Bill's midrib, as tightly as he could. Sure he wouldn't bleed out, he went from agent to agent, trying to help them as much as he could. All through it all, he prayed no car exploded or caught on fire. After all, fire was the recipe for the ultimate disaster. As he walked back and forth, the fatigue began to settle more on his body, and his legs trembled with each exertion that took a toll; but he couldn't stop. John was scared that once he did, someone would die. Walking back to Bill, he could hear the sirens of the emergency group, and for some reason it felt like a heavy weight had been dropped upon his shoulder. He felt the whole fatigue at once, and finally, his knees buckled, causing him to drop to the ground.

From the ground, he watched as the response unfolded; their synchronized movements bewitching the longer he stared. John wasn't aware of him leaning backwards. He wasn't aware of him hitting the ground. Neither was he aware of closing his eyes, and finally giving into the oblivion; the comforting darkness that comforted his soul, and cradled his tired body.

Blinding running, blindly forgetting about every other thing, and fiercely praying that everything you've heard was a lie. Jordan couldn't hear a thing; neither could see anything beyond his unconscious son. As he stood by the doorway, he felt his heart drop to the bottom of his stomach, his head light as he stepped backward; almost to reject the scene he was privy to. Different tubes were connected to Bill's body, his face pale and swollen. It was a far cry from the cheerful boy he knew. "No...No..." He looked around dazedly, his knuckle tightening against the door frame; trying to hold himself up.

Leaning against the door, the first sob broke out; a low guttural sound that almost sounded like a gasp. His lips quivered badly, his hands unsteady as he looked towards the direction his son was. Shaking his head, Jordan bit hard on his inner cheek, not stopping until he tasted blood. Breathing in deeply, he pulled away from the frame, and walked inside the room to where his son slept. On a closer look at his face, Jordan couldn't help but cry out. His already broken heart broke more, and he could see the extent of the damage done. Bill could only breathe through the machine hooked to his body.

Jordan could see various cuts by the side of his son's face; a deep crease between Bill's brows as he struggled to hold on.

"Billy boy." He cried, lightly caressing the side of his face. "I am sorry… I am really sorry." He pleaded, shaking his head as he spoke. Seconds ticked by as Jordan silently wept, his shoulders shaking as he stood in the presence of the one that b

"Mr. Martins…" A voice from behind broke into his thoughts, causing him to whip around. By the doorway was the agent that had promised his son safety; the very one that told him everything would be alright. The very one he had entrusted his son to. Anger pulsed through his blood; the strong feeling of hostility and antagonism towards the man with an unreadable expression on his face.

"Can I talk to you for a minute?" The agent began.

Looking at Bill, Jordan turned around and walked to where the agent stood. Tightening his palms into hard fists, Jordan looked down, trying to control his rage the best he could. He didn't say a thing, and he realized he couldn't say a thing at the moment.

"We are sorry for what happened to your son. We are currently launching a full investigation with the FBI to find the attackers of today's incident. Your son will be under protection 24/7; and nothing like what happened earlier would repeat itself."

"I trusted you…" Jordan whispered, raising his gaze from the floor.

"We are trying to –"

“I trusted you!” He yelled, slamming his fist against the wall. His voice echoed across the hallway, bouncing against the walls, and transmitting the anger to the world. “My son… Bill is not moving. How can someone so bright, and… and healthy become this way?” He yelled. “He was shot three times. You promised me maximum security, and he was shot three times!” He repeated, shaking his head in resentment. “I can no longer trust you… or anything you guys say. To0 you, he might just be a tool, or a job. To me –” Jordan said, beating hard on his chest. “To me he is my son, my family. My only f- family” He repeated, his voice breaking on the last word.

CHAPTER TWENTY FIVE

John wasn't aware of how much time had passed. Perhaps eight hours? He wondered, pressing the ice pack to his head. He had been lucky; or so he was told. He was the only one that had scathed through with minor issues. Lowering his hand down, he stared at his red palms; a result of the ice he was holding. They had taken Bill away from him, and he had not seen him since then. The doctor had come back to tell him the operation was successful, but that was all he knew. He was scared. Scared that something would go wrong, and Bill would… die. Closing his eyes tightly, he tried driving the thoughts out of his head, but they continued to remain, making no difference.

"Noooo!" Someone shrieked, the distressed voice breaking into his thoughts, causing him to open his eyes. Looking down the hallway, he noticed a woman dropping onto her knees and holding on to her chest as she fell to the ground. He didn't know what the doctor had said, but her whole being radiated sorrow, and he could feel his throat clog as he listened to her heart wrenching cries. "You promised me Nar… you promised." At the mention of the name of the man he was temporarily acquainted to, John felt his heart grow cold once more.

Was the man dead? He wondered.

His heart began to beat erratically, and the more he heard her sobs, the sicker he felt. Jumping up to his feet, the ice pack fell to the ground, but John could care less. At that moment, all he wanted to do was empty out his stomach. Using the wall for support, he pressed his left hand against his mouth, biting down on his lip to keep the bile away. As soon as he found a restroom, he let go of the restraint, heaving as he emptied out his stomach. This continued for a minute, until finally there was nothing left to puke. Sliding to the floor, he raised his knees up, placing his head on them. He wanted to cry, but then again – what would that solve?

He felt empty, cold and alone. He wondered if this how Bill felt when he was gone. He wondered how Bill had been able to withstand the pressure the first time. He thought about a lot of things, unaware of the time passing fast. By the time he left the restroom, an hour had gone; and that shocked him more than he thought he would. As he hurried back to the room he was assigned to, he came face to face with police men patrolling the halls. Immediately pone of the officers caught sight of him, he yelled.

"There he is." The man yelled, drawing the attention of every other officer. As soon as he was spotted, John watched something strange happen. The tense situation was almost comical as one of the police officers strode towards him. "Are you Johnny Jones?"

"Yeah…?" John responded, confusion etched on his face. Everything that happened next was a flurry of activity that left John light headed. He was ushered into an empty room, and once again he was met with the agents he had met in school a while ago.

"Mr. Jones, can I ask where you were?" The first agent asked, running a hand through his hair. This was the first time John had seen him look this distressed. The other time they had met, his unreadable expression had remained the same throughout.

"I was in the restroom."

"For an hour?" The agent asked, raising a brow. "I don't think you understand how dangerous moving around is for you."

"I was in…" John trailed off, knowing there was no use arguing. The man in front of him didn't want to hear a thing. "I am sorry." He said instead.

Sighing, the agent finally let it go. John folded his arms against his chest feeling resentful. He was angry, but he wasn't sure who he was angry at. "Mrs. Jones was here to see you."

"My mom?" John asked, incredulously.

"Yes, your mom. She was here an hour ago, but was asked to leave since we thought she might be in danger." The agent replied.

John looked down, knowing he had missed the chance to meet his mom one more time. He didn't know how he felt about that, and he wasn't sure if she was ever going to come back. "Did she leave any message?" He asked, his nails biting deeply into his palm.

"She left a note." The agent replied, handing over a small piece of paper. On it was the familiar handwriting belonging to his mom.

'Bill is fine. I am fine. Please be fine.' The message read.

Although it was short, it contained all the information he needed to know, and he found himself overcome with gratitude. His head was bent low, and as his mind was a myriad of emotions. He wasn't sure when the agents left, but he had enough time to think about everything. Later that night, as he laid on the hospital bed assigned to him, he wondered how long he was going to stay here. Although everything had happened so fast, it seemed like the main focus of why they had to leave had been postponed. Those that disappeared were still missing, and there was the risk of more people getting sucked into the loop of time. Bill was still unconscious, and John had no fucking clue on what to do, or what plan the CIA had. They never told him anything, and it almost felt like no one was taking anything serious. "For some reason I feel like I am still stuck in time." He whispered to himself, staring at the white ceiling above. As he focused on the lines above, the door to the room opened, drawing his attention towards that direction.

"How do you feel?" The doctor asked, taking his vitals as he asked some questions. "Do you feel nauseous?"

"I don't know." He replied honestly, knowing there was nothing else he could say. "And no, I don't feel nauseous."

"Okay… Let me know if there is anything that hurts." The doctor said, about to leave.

"Doctor." John called out, stopping the man from leaving. "Do you know a Billy Martin? He was brought in earlier this morning… uhm, gunshot victim?"

Shaking his head, the doctor replied. "I am sorry, but I am not the doctor in charge of that particular patient." The doctor replied.

"Oh..." John responded, unable to keep the disappointment from his voice. "Okay thanks." He added, forcing a smile.

"Alright, have a good night."

"I don't think I can sleep." John mumbled, turning his face away. Moments after the doctor had left, all John had was his thoughts, the darkness, and the silence that echoed so loud. He knew once he closed his eyes, the nightmares would keep rolling in; so he kept his eyes open, zeroing on one spot on the ceiling. But the fatigue was too much for his body to bare; and against all odds, his heavy eyes finally closed, slipping into oblivion.

Seconds rolled into minutes, and a soft snore filled the room. Nothing seemed amiss, and everything felt peaceful. But then, the eyes of the one who slept flew open in an instant. A deep crease was between his brows, his eyes heavy and gritty as he stared into the darkness. John was not sure about how it happened, but the next thing he was aware of was the disconcerted feeling you get when you've been woken from sleep quickly. A glance at the clock tells him that twenty minutes have gone by. Yet the room is still silent, with no noise to rouse him from what had obviously been a deep sleep.

Slowly, he stood up, the bed dipping as his weight adjusted on the mattress as he got off. He stood up, scratched his nape, before finally his eyes connected with the mirror on the adjacent side. His eyes were still dazed and full of sleep, his mind still confused from why he had woken up. As he continued to stare at the mirror, his mind began to clear up, and slowly he realized that his face was not the only figure reflected on its polished surface. The man behind seemed as shocked as John as they made eye contact. Full second elapses before a strangled cry escaped through his lightly parted lips, and his survival instinct kicked into action.

Yelling on the top of his voice, he turned the opposite direction, running towards the door. But the assailant wasn't having it. Reaching out, he grabbed him by the collar, yanking him backwards. The force made him off balance, causing him to slam onto the edge of the bed, and down to the floor with a thud. As his back connected to the hard floor, the air left his lungs, tears stinging his eyes from the pain. "Fuck." He grunted. His left wrist also felt painful, and he wondered if he had broken it.

Yelling once more, John wondered where all the security the agent had boasted of were. As the assailant launched for him once more, John knew he couldn't stay down. Scrambling to his feet, he reached out for anything he could grab; throwing a small vase towards the masked man direction. But it didn't hit the target. But it was enough distraction for him to attempt to slip by once more. And this time he was successful. As he burst out through the door, he realized there was no one around, and it was oddly silent. "Help!" He yelled again, but all he heard was the sound of his own voice echoing back. "Shit, shit, shit… don't fucking trust the government." He swore, heading towards the elevator.

He could hear the sound of the assailant's footstep getting nearer with each passing second, but he couldn't afford to wait. Repeatedly pressing the open button on the elevator, John wondered if he could make it. He was almost about to give up, when the elevator opened. But his relief was short lived as the assailant pushed him into the elevator, following behind.

Adrenal filled John tightened his grip around the assailant in a bid to hold him down, but the assailant wasn't having any of that. He walks back fast, slamming John's back to the elevator wall, which knocked the air out of his lungs; that one move- loosened John's arm around the assailant since he was hurt. As he slid to the elevator's floor, John grunted in pain, clenching his teeth in frustration as his side throbbed from the pain. John knew he has no chance, since it was a fight between an expert and an amateur, but he needed to save more time for the Agent to get here.

"I am not giving up so easily bitch." John swore scrambling to his feet as the assailant kicks his jaw, knocking him back down. The assailant moves closer to John intending to do more damage, but it was the chance John needed. Taking the risk, he lurched at the assailant, grabbing on to his leg. Before the assailant could kick him off, John sank his teeth into it, causing the assailant to groan out in pain.

It was a battle of survival.

John refused to die this way. No matter how many hits he got, he knew he would never let go, and would die trying. Just as the assailant was about to punch him across the face, the door of the elevator opened and by the other side was a wall of armed men. Red laser gun lights swarm around and fix on John and the assailant. His fist frozen up in the air, John cussed beneath his breath as he realized he wouldn't be able to make it with the beams pointed at him. A single shot could be fatal and that meant death.

"So this is where you are." One of the armed men commented.

Glancing to where the assailant stood, and then back at the men holding laser guns towards them, John put his hands down. “About God damn time…” he cussed again, relief washing over him. It was the Military Police with assault rifles; the agents that had been missing all along.

CHAPTER TWENTY SIX

John kept staring at the agent in front of him, his expression unreadable as he kept staring. He had a lot to say, but instead he kept quiet.

"As I said before, I apologize."

"You all disappeared." John accused, wondering if there was really anyone he could trust. "Who was that man? Why are people trying to hurt us?" He queried.

"They want both of you dead." The agent explained in a normal voice, his expression the same. Those words sent chills down John's spine, and he couldn't help but sit back.

"Who are we dealing with?" He asked.

"The onion syndicate." The agent replied.

"Who are they, and what do they want from us?" John asked, unable to keep the fear from his voice.

"What you can give them."

"The app?"

"The app." The agent confirmed, causing the knots in John's belly to tighten. "If they know so much about us, how come they don't know the app is no longer controllable?"

"I am sure they do. But the power that comes with having a key is attractive. So they don't care. I think I need to tell you about them." The agent began. "Drugs, illegal fire arms, sex trafficking, kidnapping, and every dirty crime you can think of; they have a part in it. The onion syndicate is a ring shaped organization. And..." the agent trailed off, placing both of his hands on the table. "Even with the power of the CIA, we have not been able to penetrate the first layer of the ring. All we ever got are tails that are easily cut. Tails that know nothing of the organization they belong to – human sacrifices." He said maintaining eye contact with John.

"The man that attacked me earlier... what happens to him now?"

"He has been placed in secure custody, we would try to get as much as we can from him. But I doubt if there would be any real information gained. He is probably one of those human sacrifices. "He responded. "That's the type of syndicate they are. They are a flexible organization, that collapses and disappears if need be. They leave no trail, they are cruel, they are merciless, and they are out for.

"Us..." John completed, his palms getting clammy as he clasp them tighter. "Is there a way to get away from them?"

"Death." The agent replied.

Someone once quoted, bad things happen in threes; and the worst of them was heading their way. A young female, coolly fashionably dressed, hurried down the hallway, trying to reach her teammates so she could deliver the important piece of information she had just received. As she strode down the hallway, the only thing she could hear at that moment was the sound of her heart throbbing against her chest; echoing loudly in her ears. Flinging the door open, she burst into the room, anxious to tell her colleague the news she had gotten. The forceful way the door opened startled both occupants of the room, drawing their attention to the doorway.

"New information." She said, closing the door behind her. "We have new directives from the white house." She stated. She looked at her colleague then at John, and then back at her colleague. "There have been a new set of disappearances." She started.

"How many?"

"1.5 million worldwide."

CHAPTER TWENTY SEVEN

You don't really know what someone is going through unless you've experienced yourself. The first emotion that pushed through Johnny's mind was shock; the utmost disbelief at hearing news that seemed almost too uncertain to be true.

"1.5 million?" He repeated laughing tersely. "Nah… you're kidding right?" He queried, his brows drawn so high; it showed the strain his mind was currently going through. But the woman that came bearing the news wasn't smiling. Her lips were thinned into a grim line; her face frozen and expressionless. Fear wrapped around John's body. He felt it deep within his gut. He felt it so deep, his stomach twisted into tight knots. Shaking his head as if to repel the words he just heard, he leaned back. "Nah… there's no way." He muttered.

"It is an emergency situation. Mr. Jones needs to be transported to the new facility as soon as possible." She added. Upon hearing his name, John looked at her, wondering what she was talking about. "I thought the location of the facility was breached?"

"We will take care of that."

Johnny folded his arms against his midrib, feeling cold, and for the first time – defiant. "I don't want to." He replied.

"What?" The woman

“I don’t want to go to any facility.” John repeated his voice louder than before. “I don’t trust you…none of you. I keep getting into these situations… close to death every single time.”

The woman walked closer to where he stood, only stopping within an air breathe of John. “I just told you 1.5 million people disappeared worldwide. And you decide it is now a good time to throw a tantrum?” She queried, her voice so low it sounded like a threat. “Thank God every single day that you have the minor law protecting your ass. If it was up to me, I would reveal your identity to the public; about what you have done. Do you know the amount of families out there with their loved ones gone? No explanation, no nothing. This isn’t about you going or not. You need to leave for the facility as soon as possible. Our technicians are racing against time to stop the atrocity you created.”

“We didn’t mean to.”

“Whether you meant it or not, it was created, and now you need to fix it.” The woman said firmly, her eyes drilling holes into John’s face. Few moments filled with tension radiated through the room, until finally – the woman’s phone buzzed. “We need to leave in ten minutes, I will be waiting outside.” She said. And just like the storm she created, she left in the same version.

With the door firmly shut, John stood rooted to that spot, his mind going through fifty shades of confliction.

“Don’t be mad at Alicia, She was close to Nar.” The agent explained, tilting his chin towards the direction she left. “He was almost like a father to her.” He added.

Learning about the new information didn't make John feel better; he was reminded once more that he could have been the one to die. Nar had taken the brunt of the crash; shielding the most damage away from him; and with John being in the middle, he had been cushioned by both agents on both sides. His palms tightened into fists, and his eyes burned from the tears that refused to fall.

He was just a normal sixteen year old. He didn't want this. He wanted none of this. He was supposed to be having the best time of his life, going on dates with his girlfriend; and yet – here he was the most miserable teen out there.

"We need to go buddy." The agent said, patting him by the shoulder. Almost instinctively, John shrugged; shaking the agent's hand off his shoulder.

"What about Bill?" John asked, averting his gaze.

"He would be coming with us." He replied, adjusting his tie.

"Is he awake?" John asked, his eyes widening a bit as he waited for the news that sent his heart into overdrive.

"No…" But it doesn't really matter at this point."

No one truly cares except for family. That line couldn't be any truer. John stared at the car in front of him, his mind going haywire as he stared at it. "We are traveling the same way?" He asked, his words sounding more like a statement than a question.

"Well, it is unfortunate that it isn't a princess carriage; get in the car Cinderella." A voice came from behind, edging him into the confines of the vehicle. As soon as he got in, the memories he tried so hard to forget came rushing back;' and almost immediately he broke into a cold sweat. His face was as pale as thin sheets of snow, and his breath sounded like he had run a far distance. The sound of the crash kept replaying in his head; the crunching sound of glass, the stringent smell of acid, the metallic scent of blood, the confusion, the pain; they all came rushing back, sending his heart into overdrive once more. He knew that there was no way he would ever be able to look at a car the same way. Leaning backward, and hoping the airflow would help dispel his anxiety; but nothing worked.

"Help…" He muttered, gasping as he tried to hold on to something… anything. But this time, he was riding with his least favorite person – the woman from earlier. There was no comfort anywhere.

"You got a problem?" She asked raising a brow.

"I don't feel so good." John responded, pressing a hand to his chest; as if to stop the heartburn that followed. But nothing he did helped, and nothing he said made them stop.

"As long as you don't die." She responded before turning her head away; looking outside the window. She was seating in the same position as Nar, and John wondered if she sat down there knowing that.

CHAPTER TWENTY EIGHT

Key adjusted his tie for the thousandth time; feeling uncomfortable and out of place in the conference he was currently seated. If he could describe the moment, he felt like a fish out of water. There was nothing more that described it best. Clearing his throat, the sound was louder than he expected it to be; drawing a few attentions his way. The man seated next to him smiled, and by following the actions; he unknowingly opened the gate for more conversations.

"What country ambassador?" The man asked, tilting his head to the side as he waited for an answer.

"Oh no, I am not an ambassador." Key rushed to explain, putting his hands out as he nervously laughed. "I am a FBI agent."

"The Chief General?"

"Oh no, no." Key said, scratching his nape. He wondered when all the questions would stop; but the confusion he saw on the man's face showed he didn't understand a thing. Key didn't understand anything either. He didn't know why he was here; amidst world leaders.

He wasn't sure if he had gotten a new promotion by being here, or if he had been sent since no one wanted to listen in to a boring lecture. Soon enough, the conference started, and Key became privy to information that was considered very dangerous.

It is strange but true that the most important turning points of life often came at the most unexpected times, and the most unexpected ways. Some things are so unexpected that no one is prepared for them. An example of such unexpected moments was knowing that the world could end at any moment. It was not a movie, not a novel; neither was it a joke. It was an international emergency that left everyone in the state of panic. Key felt his face turn pale the more he listened to firsthand information on how the world could cease to exist in a minute.. It felt ridiculous, but the unmistakably fear was apparent; a fear that left your insides upside down, while cold sweat broke out on your back. The fear that you could lose everything in a blink of an eye, and all you knew would cease to be.

Outside the National Intelligence Bureau; Headquarters, the sun stood gloriously above, shining brightly with its golden rays tinting the light blue sky that spanned as far as the eye could see; a very good day if one could say. Though the weather was great, it did nothing to reflect the seriousness that went on inside the building.

"First of all. I need to say this to show how important these meetings are. As of 4:00am this morning, it was confirmed that the president of Russia was found missing." At his declaration, the buzz amidst the crowd went wild, and the panic felt more real than ever. Everyone was now aware that they could be the next Malaysia President.

Key frowned at the host words, his brows drawn together as he tried to make sense of what was going on. But no one was listening. Everyone was absorbed in their own world of interpretations.

"If you look at the screen, you can see the amount of people that disappeared yesterday, and if you check the second image you would see how far it has travelled in just one day." The host pointed out, circling the projected images with a light pen. "With every minute that passes by, we are constantly faced with the threat of being the next person to disappear."

"What has been done so far? What can we do to stop the disappearances?" A representative from Nigeria asked. "I thought the last meeting you said something would be done? How come there is no change but it has gotten worse?"

"The measures applied included using the CELO elite hackers. They tried hacking into the subsequent database, but that had little to no effect."

"CELO? Isn't that the international team that's so hard to contact. And they were able to do nothing?" The representative from Malaysia asked.

"Unfortunately, there has been no success. We also tried medium-distance collision, by creating new magnitude wavelength around the satellite that transmitted it across... but as you can tell, that didn't work too."

"I heard there is another option to hacking into the database? Something that has to do with the real inventors; is there a way to know their status progress?" Key said, speaking for the first time since he came. His question drew in the right amount of attention, and although most people couldn't recognize who he was, they all waited for an answer that held much hope.

"We are currently on it, and as we speak we would be going live in the next 48hrs. It will be a chance for the other countries to see what we have been doing so far." The administrator explained, the slide on the projector changing to a different image. "The only problem is, unfortunately one of the key components was attacked. And is currently indispose... we don't know yet how much it would affect the whole operation; but we are hoping for the best."

The last piece of hope everyone was tightly holding on to disappeared like sand in a tight fist. Key closed his eyes tightly, trying to steady his heart that was beating erratically. He could hear the panic shouts, he could hear the administrator trying to calm the crowd, he could hear the low sound of an AC across the room; he could hear it all. But nothing mattered. All that mattered at that moment was the heavy truth that was hovering above their head. Opening his eyes, Key squinted against the bright lights as he faced the direction the administrator stood. "What will happen when it finally hits?" he asked, but amidst the loud voices, his words were lost. "What will happen?" He asked again, but nothing changed. Feeling frustration from deep within, "What will happen when it finally hits again?" Key yelled, the microphone amplifying his words, echoing and bouncing it across the room.

As his words echoed across the room, the noise stopped, and everyone stood silent for an answer that would shatter the temporary peace. "What will happen when it finally hits?" Key repeated once more, but this time in a lower voice. Everyone changed attention from him to the one who held all the answers.

"The worst case scenario that would happen with the loop in time includes; the disappearance of every single person on earth. Earth would simply not exist, and if the planet doesn't exist – the people won't" He replied in a somber voice, scanning the expressions of everyone in the room. "If that should happen… it would lead to the death of every single living being or thing on earth."

Since the beginning of the meeting, this was the first time the silence was so loud. It was so loud that even if a pin was dropped, it would echo loudly. The stark reality of what was to happen was before them; a reality that placed them between the devil and the edge of a cliff. The reality that they would cease to exist just as the dinosaurs had been millions of years ago.

"Billions of dollars of research and this is the result you bring to us."

The host administrator pursed his lips, a visible vein popping by the side of his neck. "We are trying our best…" He started once more, only to be cut short by another official.

"Trying? A repeat of what happened a million years ago would happen, and that is the same as everyone going extinct. We don't want trying, we want result." The Canadian ambassador said.

"That's why we are here today to discuss –"

"Did you guys even try to stop it?" The representative from Brazil yelled, cutting his words short once more.

"What of the nuclear effect? Can't more nuclear bombs be released?" The representative from Yemen asked, banging his hand on the table.

Key knew most of the people seated here were politicians that had no relation to science, so it was quite expected that such questions would pop up; but none of their scream could solve anything. He had seen a lot of their kind throughout his early career, when he was still in the FBI as an Intelligence Agent. But at this point, the voices grew louder; showing more discontent and fear as panic radiated throughout the entire room. All through this, Key said nothing. His mind was no longer in the room, but had traveled far beyond it. He was thinking of what the world ending really meant.

"The only way to eliminate the threat of the looping time hitting the earth is to – evacuate." The Administrator said. Shoulders stooped, head bowed, he removed his glasses. "We leave earth." He muttered.

"Do you have the resources to evacuate the entire population?" The Britain Prime Minister asked.

"No we don't. But we have the ability to evacuate some of the population."

"How many?" He asked.

"Ten thousand."

"Where do ten thousand go?"

"Hide out in space."

Another chaos broke out.

CHAPTER TWENTY NINE

He wished he could run after him, but doing so would definitely bring untoward attention towards him. Of course, people kept throwing courtesy greetings towards him; but He didn't have the heart to stay and greet back. He walked briskly down the hallways, and hurried to the elevator; hoping to meet up with The Viper.

As the elevator rolled down, He felt a new wave of nervous fever rip through his body, and his stomach once again – began to hurt. He didn't realize he clenched his teeth so bad, and it wasn't until he breathed through his mouth that he realized it was beginning to hurt. Lowering the intensity on his jaw, He look forth; and as the elevator door pinged and opened, he strode out, and walked directly towards the building entrance; scanning the perimeter for a man that had brought back his nightmares.

Frantically searching, He sifted through the faces; but every face his eyes connected with was not the one he wanted to see. Stepping out of the building, he headed towards the parking lot; praying to God he didn't miss the chance to talk to The Viper today. What would he do if he started talking to the wrong people?

"I knew you would come after me." The Viper smirked, pushing away from where he rested against the wall; the sight of The Viper, grinning widely like a Cheshire cat irked him to no end. Once upon a time, The Viper had made sure his problems went away like ash in the wind; cleaning it all up like they had never existed. The Viper had been damn good at his job, erasing any connection He had with the onion ring; every single one of it. But little did he know that it was all a front. Once you were involved with the ring, you were involved with it for life.

"How's Mia." The Viper greeted. At the mention of his daughter's name, he felt the veins pop from his forehead. The only reason he was here was because he owed The Viper a great deal, but seeing him again was not what he wanted; his reappearance had made him forget the good news as his heart sank further in his belly. He had never thought he would be staring into The Viper's face ever again. The sentences '*don't count your hens before the eggs hatch*' floated through his mind.

"What are you doing here?" he snapped in harsh whispers, looking around in fear that someone would see them both. "Let's go somewhere else." He said, his voice leaving no room for disagreement as he walked away; not looking back, but expecting The Viper to follow. He led him to a more discrete place, his heart quickening with each step they took away. Immediately The Viper was in the enclosure, He turned at him. "Why are you here?" he asked sharply.

"Hello to you too…" The Viper scoffed an amused grin on his face.

"How did you find me?" He asked again, trying to change tactics.

"Easy there man, easy okay." The Viper laughed. "Stop trying to rack your brain over how I found you? You forgot who I am? I am fucking The Viper Leroy. There is no fucking place in this city that I don't get my eye on. My eyes are all-seeing in this place." He boasted.

"Why now? Why are you looking for me now?" He asked. "And how did you really find me?"

"Seriously? You still on that shit? I mean… who told you to be involved in this case. The big boss is interested, and that's all you need to know. . I am really proud of what you achieved, really. I mean, we go way back to back Yo; you know what I am saying? Especially every time I saw your face in any case connected to what I am walking for' I am always so happy.

He eyed The Viper's expression, not getting fooled for even one minute. So his getting a job with the FBI had come with a price? But the question now is – what comes next. "What do you want from me?" he asked.

"Hey – I am not that type of guy!" The Viper protested, raising his palms up in the air between them. "What's wrong with a good reunion? Everyone loves a good reunion, right?"

"Cut the shit out, what do you want?" He repeated, seething.

The smile faded off The Viper's face, and the serious expression he was known for came on. "I need to know the next plans taken; to the last details."

He felt sick to the gut, and though he had been expecting this from the very start – it didn't make him feel any better.

“It is simple, I got a proposition.” The Viper continued. “I need more proficient information, and then we are good.”

He narrowed his eyes at The Viper’s words; there was no way it was that simple. “What else do you need?” He queried.

CHAPTER THIRTY

The beeping sound of a machine filled the air, and it didn't take long to know what was causing it. It was the life machine that Bill was currently hooked on. The hospital room was in pristine condition, with the underlying scent of antiseptic lingering in the air. Outside the hospital window, it was the same world, but very different. The insects buzzing around were not the only thing taking time off; Jordan had also decided it was time for him to take a little break. There was nothing better than good health; and even though he had not been sleeping well lately, he tried catching up on lost sleep. But things were not going as he planned. The insistent ringing of his phone jarred Jordan from his slumber, causing him to groan as the sound continued to penetrate his subconscious. He rolled over in bed not fully awake as he picked up the call without checking who it was.

"Hello," he said groggily, slowly slipping back into oblivion before his mind caught up with him, and dragged him back to the present. He looked out his window, squinting his eyes to see what time of the day it was; he saw that it was still pretty early, causing him to cuss under his breath. He hoped the person calling had better be someone important.

"Good morning, it's me Marcus. I'm so sorry to disturb you but I really need to tell you something." The voice rang out. Jordan knew who that was; it was Marcus, his longtime friend, and the urgency in his voice made Jordan more alert. His first thought was on Bill, and through squinted eyes he stared at his son. Nothing seemed different, and he could visibly relax as he realized it wasn't about him.

"Marcus? What's going on? Is something wrong?" Jordan asked, sitting up. He ran over what could possibly be wrong and the series of incidents that had happened recently coursed through his mind; causing his muscles to grow tense as his mind kept racing through each possibility.

"Do you remember that I told you that I was working on something last week… Do you remember?" Marcus asked, pausing to let him jog his memory, even though he knew Jordan would not forget.

"Ye… yeah..." He responded, clearing his throat that was still thick from sleep. "What about it? You never told me what it was?" He added. One of Jordan's many talents, aside from being a good father, was his ability to store a lot of information in his mind; it was one of his greatest strength, and then again – a flaw. Just as good memories stuck; bad memories also stuck.

"Congratulations…" Marcus burst out, causing Jordan to be confused and puzzled about what this was. But one thing was for sure, it was evident that it was something good when comparing the high pitch tune of Marcus to his normal voice.

"Uhm… thanks I guess. Is there something you need to share with me?

"Of course…" Marcus scoffed. "And you owe me one; I worked my ass off for this."

“Hmm…” he muttered, his heavy hooded eyes closing and opening. “Marcus, can you get to it. I am really, really, busy at the moment.”

“Sorry, sorry. I got too excited and forgot myself. Well – brace for it, wait for it. Jordan Martin, I have successful found a way to retrace IDI syntheses. A

There was a moment’s silence of Marcus’ announcement, and then his voice sounded down the line.

“Jordan! Isn’t it wonderful?’ he said tremulously. “I am finally able to copy the syntheses code! I finally made it!” He repeated.

Jordan couldn’t believe his ears; happiness, joy, bewilderedness, delight ran through him leaving him speechless. His once heavy eyes were now clear, and the sleep was long gone.

“Jordan?” Marcus asked, wondering if he still on the phone.

“I will call you back in a minute.” Jordan replied gruffly, cutting the call.

He was filled with gratitude; gratitude that clogged ones throat and left you with a painful sensation in your throat as you fought the tears. At that moment, the sound of the machines went crazy; draining the happiness Jordan felt at that time. His first instinct was to run towards the bed, and although he couldn’t do anything, his body was already next to his son.

The sound of the machines began to blare loudly, and each sound was like a gunshot to Jordan’s heart. Reaching over the bed, he pressed the emergency code, praying the doctor came just in time. Reaching for Bill's hand Jordan held on tightly. "We are just getting started buddy… come on… please. Do it for me! Come on!" He pleaded fervently, but life was a shitty place. The machines stopped the same time the doors burst open.

There was one person that was long forgotten. Left with a lot of questions, with no answers in sight. She stared at the sun through the gaps in the leaves; sparkling like stars in the night stars, and dazzling her eyes with the bright lights. A gentle breeze blew inward at that moment, gently ruffling her hair, and teasing her nape. "It's strange…" She said, staring at her phone screen for the umpteenth time. "It's just freaking strange." She repeated.

"That's your favorite word these days." Her friend replied, shaking her head.

"Hear me out… two boys suddenly disappeared overnight, and so did their families. No one knows where they went, and no one saw them since the moment they disappeared. It doesn't make any sense." She grumbled.

"It is quite strange." Her friend agreed.

"That's what I was telling you all along!" Rita grunted, turning away. Although she didn't know John for a long time, she knew he was not the type to disappear without a word. She remembered the last time she saw him as clear as night. They had plans after school, but John never showed up. She remembered calling his phone repeatedly, and getting mad; thinking he had ditched her. But John didn't show up the next day, or the days following that. His presence had been erased, and she realized how little she knew of him.

"Rita..." Her best friend began, hesitation in her voice as she called out to her.

"What?" Rita asked, turning to look at her.

"The recent disappearances... you don't think..." She trailed off, unable to complete the sentence.

"Nah..." Rita laughed nervously, turning away. It would be a lie if she said she hadn't thought about that scenario more than once; but then, her mind simply refused to believe it. There was a lot of conspiracies online about the disappearances, but not one held any answer on how to bring them back. People were randomly disappearing; a mysterious pattern that had no explanation. The amount of people that disappeared continue to increase each day, and that made her more scared than ever. *Was she next? Or her family? Or her friends or a random stranger across the street?* Those thoughts ploughed her mind.

"I am not saying I am praying for that to happen to John." Her friend rushed in, trying to explain herself.

"I get it... it's okay." Rita replied, wetting her lip with the tip of her tongue. "We should leave." She added, standing up from the bench she sat. As they walked away, Rita tried to stop thinking about him; but then again – she couldn't help it. Long after her friend had gone, Rita doubled back, heading to a place she had visited a lot of times – she was heading to John's house.

The path was now a familiar place, and it didn't take long for Rita to reach the empty house. But almost immediately, she realized something was off. Someone was in the yard, and the thought that it could be John sent her heart racing. "Hello? Is anyone here! Hello!" She yelled out, hoping to draw the person's attention. Few minutes later, someone showed up, but it wasn't the person she wanted to see. He was a middle aged man, and Rita wondered if he was someone related to them. "Good afternoon Sir…" She greeted.

"Yes… Good afternoon." He greeted back, his hands akimbo. "How can I help you?" He asked with a smile.

"Yeah uhm…" She rushed out, pulling out a picture of John on her phone. "Do you know a Johnny Jones?"

"Johnny Jones?" He repeated, squinting as he stared at the phone screen. "Nope. I don't know someone with that name. Perhaps the old owners of the house. I am the new owner of the house." The man responded, returning the phone back to her.

"New owner? That's not possible. They were still here last week." Rita argued.

"According to the realtor, the house has been empty for a month." The man replied. "I just moved in today." He shrugged

"What?" Rita muttered, unable to believe her ears. *'How was that possible?'*

"You might probably ask the neighbors… I don't think they just moved in." The man suggested.

Rita bit down on her lower lip, unable to answer. She had already done that. Asking a ton of questions to the neighbors on both sides and across the street. No one had seen them, neither did anyone hear from them. “Do you have the contact address of the former owners? Or a number….? Anything?” She pressed on; a last stretch in hopes for a favorable answer.

“I don’t know anything about them.” The new owner shrugged. “I am sorry about that.” He said apologetically.

“Nah… it’s okay.” Rita replied, forcing a smile. “Thanks for your time.” She added, walking away. *‘John where are you?’* She thought.

CHAPTER THIRTY ONE

The ones who ruled the world are often those we overlook. Once the doors to the dam are open, the water becomes unstoppable; bursting through its restraints and creating havoc on the places it can reach.

The president of Russia was that man. Not only was he the head of the country, he was also a man who had his hands in a lot of pies; Pies that could never be revealed to the world. The black network had their own source of information, and it didn't take long for them to find out that truth.

"I guess it's time." The man seated in the darkness muttered, drumming his hand by the rim of the table. His fingertips created a rhythmic melody, a backdrop to the disaster yet to come. "Call razor." He muttered, resting backwards. With the weight of his back against the back of the chair, the wheels under shifted away, pushing him away from the table.

He took the opportunity to turn away finally, the darkness swallowing his whole figure. The soft thud behind signified the door being close, and once more he is left alone in the shadows.

Unbeknownst to the world, a conspiracy so dark was about to take place. The darkness saw an opportunity to sneak into the light, and none were willing to let this slide. It was their chance to regain the things they had once lost, and they were starting with one of the most important sectors.

John grabbed on to both sides of his hair, wishing he could pull his hair out. Not only was he frantic about the new development, he was forever conscious of the invisible burden that weighed heavily on his shoulders. "God… I don't… what should I do now…" He mumbled, his words inaudible to whoever try to listen in.

Finally raising his head, he scanned the new conference room, his mind cracking as he stared at the hundreds of people before him. On each walls were big screens, showing different information, with different figures that were constantly changing. Lowering his gaze, his eyes settled on the people walking the floors; the combination of both resulted in total chaos. The hum of the machines, the buzz from the humans, and the thick tension that laid between. They were expecting him to solve a problem he didn't know, and each time someone looked his way, his heart skipped a beat. *'What should I do?'* he thought, the beginning of a headache creeping up his temples.

No one that stood below could understand what was going through his mind. He was scared, he was nervous, he was filled with angst. His mind was filled with a myriad of emotions. Clasping his hands together, he considered getting and running away; as far as his legs could take him. But he knew the dangers of doing that. At that moment, John felt totally alone, and terrified at the same time. 'Bill… was this how you felt?" He muttered, his breathe shaky as he said his thoughts loud. The sound of footsteps drawing near made him left, and few steps away was a man he had never seen. For those seconds, they shared a gaze, and as the awkward feeling set in, John looked away.

“Can you see that?” The man asked. But John couldn’t reply. There was suddenly a lump in his throat, stopping any sound from passing through. “One app is ruining the whole world. Sound absurd… doesn’t it?” The man continued, spreading his left hand in the air. “

Unconsciously, John followed his hand movement, his eyes scanning the large screens that showed the total number of missing in the world. “Do you know what demons you unleashed?” the man queried once more. “In one night, the power balance was tipped. In one night… a lot of things began to happen. No one told you anything, right?”

Although John wished the ground could swallow him up at that moment, he remained in the strange man’s present; unable to run, unable to speak.

“Russia's President is missing…. The financial market is crashing, innocent civilians are missing, more could go missing, the black network are running rampant… and yet-" He paused, his eyes turning into hard slits as he looked at John. “And yet, no one is blaming you.” He scoffed. “Get ready.” The man added, walking away.

Releasing a sharp gasp, John realized he had been holding his breath for quite a while. Bending over, he held on to his side, while trying to regulate his breathing. His lungs felt stuffy, and it felt like he was suffocating with each second that passed by. The noise in the background sounded amplified, and the ringing in his ears didn’t stop.

It wasn't a self-party. It wasn't a moment of dereliction. But even though the whole world had their eyes on him, the fact that everything happened due to an accident didn't change. The fact that he was still a teenager thrust into a world he didn't belong, also didn't change.

Clenching hard on his teeth, he leaned backward, staring at the ceiling for a few minutes. Something ran through his mind, and soon enough he was up on his feet. On his face was a different look; an expression of determination mixed in with a new set of motivation. He sighted the man that had walked up to him earlier, and without missing a step, he walked towards that man. He had his back towards him, and he was in the midst of several other people. "Hey." He called out, his voice low and his lips quivering for a split second. But the trembles only lasted for that long. It was almost like it was a mirage, and nothing of sort ever happened. "I have something to say." His words were enough to draw their attention, and five pair of eyes stared at him with curiosity. As their gazes bore holes into his skin, John said the words that were hidden in his heart for so long.

"I think there's something I can do."

CHAPTER THIRTY TWO

The sound of hurried footfalls filled the hallways, and soon enough it turned into longer strides; as the owner broke into a run. "Fucking arses, I wish they all break their knees and fall into a pit they can never get out from." He mumbled with absolute hatred, a look of abhorrence on his face. With enough force to take the door off its hinges, he pulled the door open; causing a squeaky sound as it scraped the wall. "Jesus!" The man behind the doors swore, a look of surprise and fear on his face as he held on to his chest. "What is going on?" James Pittsburgh asked, a crease forming between his brows.

"Don't fucking ask what's wrong. Have you fucking checked the news this morning?"

"Tone it down on the curses… what's going on?" James asked, his face morphing into a confused look. "Did something happen?"

"My God James!" The angry man berated, touching his temples as he spoke. "Don't ask me. Check the freaking news." He said in an exasperated tone, pacing as he completed his words. "You really have no idea what's going on. What do you do anyways? Why are you always so clueless…? You don't know how much damage has been done this morning." He rambled on, not stopping as he paced in front of the table.

"Alright, alright.... Give me a damn minute." James swore, pulling his phone out. As the search results came out, the expression on his face changed almost immediately, and the blood on his face bleached out in a second; leaving behind a pale face. "This can't be true."

"It is fucking true... I confirmed with Wall Street before heading here. Everything... every single thing crashed in a minute. The investments, the cattle's... it's all gone." The angry man explained, his face getting redder with each word he said; a complete opposite of James' face. James looked at his phone one more time, wondering if he was truly dreaming. The two top searches were terrible news, and it somewhat reminded him of bad luck happening in threes. The first search topic was about the mysterious disappearance of people around the world. It was no longer a rumor, but a stark reality as the true details were now out. And just below the first was a new addition that sent his blood pressure high.

The financial market had crashed overnight. Investments that were proven to bring top gains were now a skeleton of their former glory. "Where did all the money disappear to?" He whispered, not aware that he had spoken his thoughts out loud.

"I don't know James... I don't know." The angry man replied, falling into the chair opposite where his friend sat. He was wholly distraught.

James clicked on a new article that had just been released, and as his eyes scanned the details, his heart sank further.

In the blink of an eye, I have found myself uttering the words "worst since the Great Depression" far too often: the worst twelve

month job loss since the Great Depression; the worst financial crisis since the Great Depression; the worst rise in home foreclosures since the Great Depression. In my previous life, as an economic historian at Berkeley, one of the things I studied was the Great Depression. I thought it would be useful to reflect on that episode and what lessons it holds for policymakers today. In particular, what can we learn from the 1930s that will help us to end the worst recession since the Great Depression?

To start, let me point out that though the current financial crisis is unquestionably severe, it pales in comparison with what our parents and grandparents experienced in the 1930s. Last Friday's employment report showed that unemployment in the United States has reached 8.1%—a terrible number that signifies a devastating tragedy for millions of American families. But, at its worst, unemployment in the 1930s reached nearly 25%.1. And, that quarter of American workers had painfully few of the social safety nets that today help families maintain at least the essentials of life during unemployment. Likewise, following last month's revision of the GDP statistics, we know that real GDP has declined almost 2% from its peak. But, between the peak in 1929 and the trough of the great Depression in 1933, real GDP fell over 25%.

I don't give these comparisons to minimize the pain the United States economy is experiencing today, but to provide some crucial perspective. Perhaps it is the historian and the daughter in me that finds it important to pay tribute to just what truly horrific conditions the previous generation of Americans endured and eventually triumphed over. And, it is the new policymaker in me that wants to be very clear that we are doing all that we can to make sure that the word "great" never applies to the current downturn.

While what we are experiencing is less severe than the Great Depression, there are parallels that make it a useful point of comparison and a source for learning about policy responses today. Most obviously, like the Great Depression, today's downturn had its fundamental cause in the decline in asset prices and the failure or near-failure of financial institutions. In 1929, the collapse and extreme volatility of stock prices led consumers and firms to simply stop spending. In the recent episode, the collapse of housing prices and stock prices has reduced wealth and shaken confidence, and led to sharp rises in the saving rate as consumers have hunkered down in the face of greatly reduced and much more uncertain wealth.

In the 1930s, the collapse of production and wealth led to bankruptcies and the disappearance of nearly half of American financial institutions. But this time, I wonder if it has something to do with the mysterious disappearance of the financial mogul, and the president of Russia. This, in turn, had two devastating consequences: a collapse of the money supply, as stressed by Milton Friedman and Anna Schwartz, and a collapse in lending, as stressed by Ben Bernanke.

In the current episode, an outside intervention such as derivatives led to a direct relationship between asset prices and severe stress in financial institutions. Over the fall, we saw credit dry up and learned just how crucial lending is to the effective functioning of American businesses and households.

Another parallel is the worldwide nature of the decline. A key feature of the Great Depression was that virtually every industrial country experienced a severe contraction in production and a terrible rise in unemployment. This past year, there was hope that the current downturn might be mainly an American experience, and

so world demand could remain high and perhaps help pull us through. However, during the past few months, we have realized that this hope was a false one. As statistics have poured in, we have learned that Europe, Asia, and many other areas are facing declines as large, if not larger, than our own. Indeed, rather than world demand helping to hold us up, the fall in U.S. demand has had a devastating impact on export economies such as Taiwan, China, and South Korea.

The man in the shadows smirked; a sinister grin that shadowed his accomplishments. He was millions of dollars richer, and each second- more money rolled in. He reveled in how easy it had been, how easy it was to chip at the little crack in the middle. They had seen a disaster, he had seen a chance. In the background, the recorded sound of classical music played smoothly, and with each note it played, his heart hit a new crescendo. Instinctively, he raised his left hand into the air, twirling it slowly like he was directing an orchestra. He had started out low. Building and climbing the ladders of the world until he was finally at the apex. In this world, seeing the little things no one saw, and moving with your instincts sometimes paid. It was all about the risk. This time was no different. He had infiltrated the market months ago, and the result was similar to that of the total fiscal expansion in the 1930s was very small indeed. Unnoticeable, and the first step to the whole operation.

He remembered when he was much younger; a time when eating two times a day was considered a luxury. He had grown up with five siblings in a tiny apartment at the back alleys of New York City. Unlike the glamor associated with the city, the slum part was a different world entirely. With burgeoning mountains of decaying trash across fifty squalid blocks of the slum area. It was an album of rubbish choked, rat infested, crime rampant area with no hope in sight. He remembered the smells, the tears, the constant anxiety that they would evicted from the apartment the next day. One night, climbing up through the protruding pipes by the side of the building, he had climbed up the roof. In the vast distance was a thousand bright lights; twinkling and blinking, a contrast to the starlet's sky. By the time he was sixteen, he ran away from home; and not once had he ever looked back, or regretted his actions. He had vowed at that moment to become someone no else could trample; and here he was today.

Back to the present, the man continued to remain in the shadows. He picked up his lighter, and put on the cigarette. Drawing a long one, he exhaled, the smoke filling the confines of the office. As the taste of nicotine hit his nerves, he began to visibly relax, making him close his eyes. 'A nicotine stick is always the answer.'

The man in the dark reached into his pocket, pulling out one of the multiple phones he had. As soon as he touched the screen, the finger print verification lit up the screen; giving access to the content of the phone. The lights illuminated his face for a split second, but that was short-lived as he dialed a number and pressed the phone against his ear. Few seconds of silence ensued as the phone on the other line began to ring.

"I told you, didn't I?" The man in the shadows greeted, his mouth lifting into a grin; a contrast of his white teeth to the dark room.

"You should have asked me to invest more." The person replied, followed by a chuckle from both of them. It seemed like any other conversation to the third position watching; but it couldn't be more off. They were both cajoling about a matter that dealt with millions of lives and billions of dollars, a situation that would determine if the financial market would forever crash.

"I heard the President of Russia is still missing... anything with you."

"Me?" The man in the dark queried, unconsciously adjusting the tie he wore. "My hands are clean."

As soon as he said those words, the man on the other side began to laugh once more; a gasping sound that sounded like an asthma attack. "That's very funny.... Very funny." He added between the laughs.

The man in the dark remained quiet, his expression unreadable and cloaked by the darkness around him. "Nature gave me a gift, who am I to say no?" He asked.

"What If he comes back?" The man on the other line asked.

"He won't... I'll make sure of it." He assured.

The call continued for a few more minutes, before finally the call ended. Silence reigned once more in the corner of the shadows, and the man was once again left with his thoughts. While the direct effects of fiscal stimulus were small in the Great Depression, the man in the shadows thought it was important to acknowledge that there may have been an indirect effect.

It was the beginning of the end of an economy he no longer had use for. He knew they were going to fight back; to defend the gold standard and maintain the prevailing fixed exchange rate. But he had a counter for that. They were most likely going to suspend the convertibility to gold and let the dollar depreciate substantially. And later on, they would try to trade gold at the new higher price, large quantities of gold flowed into the U.S. Treasury from abroad. It was a predictable move he had seen long ago. He had after all planned this for a very long time. These gold inflows would serendipitously continue throughout the years, and he was planning to milk it for as long as he could.

He was going to watch it all unfold, including the political tensions mounting in Europe and investors searching the safety of U.S. assets.

It was a master strategy no one would expect; a well thought plan that dealt with cold hard cash in a clinical way. As soon as the music in the background stopped, he tilted his head to the side; the silence echoing in the room he stood. But with a single clap of his hands, the music started all over, and the intense mood reached a new heights. The identity of the man remained unknown, so were his features. But a fact remained – he was the main character of the new arc, and he knew he would continue to be that way for as long as he wanted.

But there was just one problem. He didn't want the ones lost to come back; most especially the most recent disappearance. The hovering threat of a solution to the disappearances gave him a concern. Walking towards the table, he picked up the most recent update on his young targets. He stared at their faces, a new found admiration sprouting in his mind. He wished to have the app in his own hands. He wished to have all the power in the world. He wished to have it all, even if it meant taking care of those that got in his way. One way or the other, he planned to get his way.

CHAPTER THIRTY THREE

He felt pain, he felt heavy, and at the same time – he never felt more alive than this moment. As his eyes slowly opened, the brightness of world caused him to wince, and in response he closed his eyes almost immediately. He stayed that way for a few more seconds, and decided to try once more. Slowly, he inched his lids wider until he could focus on something. Through heavy hooded eyes, he stared at the white image in a daze; his brain trying to process what it was. As his vision became clearer, he realized the white object was the ceiling. *'Where am I?'* he wondered, trying to scan his new environment through the corners of his eye. And the first thing he noticed was the machines on the side, filled with cables that seemed to be connected to him. *'Hospital? What happened?'* He wondered once more.

Like a camera roll, the memories from the past came flashing by, and the blank places in his memory came rushing back. He could feel the ringing in his ears as the scene of the crash came with a sharp sting. But with realization came fear. He couldn't move any part of his body except his eyes. The fear of being paralyzed sent his heart racing, and as a signal sent the machines beeping. The loud beeps increased his confusion, and the panic that came along was inevitable. Distraught, and wishing he could move a part of his body, he was unaware of the person that stood by his side.

"Bill! Son! Can you hear me? Bill!" The voice reached out to him, guiding him away from the deep pit he had dug himself. Overcoming some of the overwhelming emotions, he turned to the side, his face etched with confusion as he stared at the man that look at him with worry and love. It took a moment for the features to register, and he was filled with surprise once more.

"Dad?" He croaked, his throat feeling parched, and his mouth feeling dry.

"Yes son, it's me… it's your dad." Jordan sobbed. As he reached out to touch his son, the doors flung open and the team of doctors rushed in. Almost immediately, they got into action, taking his vitals and reading his charts. "How do you feel Bill? Can I call you Bill?" The doctor asked, monitoring his eyes with a tiny flash.

"I guess…" Bill responded, turning towards his dad; who was now at the very back. "I am very thirsty." He groaned, turning back towards the doctor. "I am thirsty." He repeated.

"No worries." The doctor assured, signaling to the intern doctor who was closest to the water pitcher. The first taste of water on Bill's mouth tasted like heaven. The cool liquid swirled down his throat, and he had to muster the strength not to mean. But like every human in the face of pleasure; they became greedy for more. "Easy… easy." The doctor said, gently guiding the cup away from the cup. "How do you feel now?" He asked.

"Much better." Bill replied, his voice clearer than before. "I can't move my body." He complained, clenching hard on his teeth as he waited for what the doctor would say.

“I guess you can’t… You were shot three times.” The doctor replied. At his words, Bill's heart rate began to pick up, and his breaths began to shorten.

“It’s nothing short of a miracle for you to be alive.” The doctor added, his grin widening. “You are a fighter Bill.”

Bill scanned the doctor’s expression, his fears still at the forefront. Although the doctor was smiling, it answered none of his questions. “Am I paralyzed?” He asked, closing his eyes to the news he was about to hear.

“Well… no you are not.” The doctor replied, causing Bill to open his eyes so fast.

“I am not?” He asked, not sure he had heard correctly.

“You are not paralyzed Bill.” The doctor assured.

“But I can’t move any of my limbs?” Bill asked, a new crease forming between his brows.

“That’s because you haven’t moved in a long time. Your joints are just stiff.”

“How long have I been in here?”

“Two weeks and three days.”

“OH…” Bill muttered, his mouth forming an O shape as he absorbed the new information. Turning his head towards his dad, he said. “Dad, they said I am not paralyzed.” He grinned, his eyes reflecting joy as he stared at his own family. But then, the joy was short lived as his brain registered one more fact it had skipped. "Wait… where is John and what happened to the world?"

The room fell silent.

"Optimus." Bill said.

"I can help." John repeated, folding his arms against his chest to hide his nervousness. The little crowd stood silent to his declaration, and John wondered if they had heard his. "I can-"

"I heard you the first time Mr. Jones." The strange man cut in, holding a hand up as he stared at John. "And what exactly do you think you can do?"

"Try to reverse the app effect." He replied, shrugging.

"And you think you can?" The man asked, raising a brow.

John wasn't sure of anything. He wasn't sure of what the next step was, and at this point – he was bluffing. There was no other word to describe it; but what could he do? He couldn't sit and watch the world go by. Bill wasn't here and he had to be strong for both of them. "I have an idea."

"Mr. Martins, I appreciate your-"

"I built that app with my friend." John said confidently, cutting short the man's words this time.

“You are proud of that fact?” The man queries, tilting his head to the side as he studied John.

“What can I say at this point? We already established that fact long ago.” He said. John couldn’t help but wince inwardly as he said those words. But there was no truer words. He had helped create the app, and that was it. “I need a laptop, I need to show you something.” John said, unfolding his arms.

A man out of the crowd stepped forward, his eyes as hard as marbles. “And do what, haven’t you done enough?”

“It’s okay.” The strange man said, patting the other man's shoulder. “Let’s give him a chance.”

The once noisy room grew silent as they gave way for the young man who was the main character for the moment.

“Here you go.” The man said, pointing to a station that was unoccupied. As soon as John took his seat, the earlier crowd stayed by the side, alongside few people. John was nervous, he couldn’t lie about that. This was the first time he was doing this without Bill; and he felt his absence stronger than ever. His mind began to fill with doubts, and butterflies began to gather down his belly. It felt like a whole was running through his veins.

“We are waiting Mr. Jones.” The familiar voice of the strange man came from behind, and John knew this was reality. This was the moment to decide his worth to the whole team. He could feel their ridicule without seeing it, and he knew he would never leave down that memory. Breathing in deeply, he began to recreate the codes that were ingrained in his memory.

He began to create a copy of the codes that had started the whole thing. The screen began to fill with codes; a beautiful array of numbers and letters, and sometimes special characters. It was a semblance of the future; an intricate construction that left a deep impression. Once again, the screens started to fill with lines of different color codes, flashing-up and racing up on the screens, and in the background Athena' thousands of blue server lights, flashed like blue stars in the night... all the display screens go black, and then a white cursor just flickered and flickered, and then the screens, changed into IT blue, and fill with racing lines of white bios code.

"The Test Post Request with Sqlmap method accepts four influences. The first influence is the URL that will be sent to Sqlmap. The second influence is the data that will be in the post body of the HTTP request—be it POST parameters or SOAP XML. The third influence is the value that will be passed in the SOAP Action header in the HTTP request. The last influence is the unique value that is vulnerable. It will be replaced with an asterisk in the data from the second influence before being sent to Sqlmap to fuzz. After we bypass the message to the screen we might be able to get a contemporary URL; and once we tested it out, it worked. So we created a Sqlmap Session and Sqlmap Manager objects." John explained.

Everyone gathered next to him knew what those codes meant without him uttering a word. They couldn't say a thing, neither could they refuse that the one they looked down on, was a genius… although he was portrayed as a villain to the world. "These are the original codes." John said, retracting his hands from the keyboard. "This should have been the result." He added.

"Then what went wrong?" Someone from the crowd asked.

“The APP gained a life of its own.”

The room grew silent once more.

CHAPTER THIRTY FOUR

Bill could feel the sweat beads on his upper lips, on his neck and by the side of his temples. He could also feel his muscles contract, and his teeth clench as hard as he tried to move from the position he was currently in. "Come on... come on..." He gritted, grunting as he tried once more. Giving up with a final gasp, he leaned backwards once more; frustration and tiredness etched on his face. Although the doctor told him he could walk, he was beginning to seriously doubt their diagnosis. Sighing, he tipped his backwards, staring at the ceiling he had first seen when he came back to the world. It was still as white as the first time; without a blemish while it glared down at the stain filled world. Nothing was going his way. Since the moment he made that app, everything had gone upside down. It was almost as if the app had sucked out the happiness he once knew.

No one was telling him anything. He wasn't aware of what was currently going on out there; and that added another layer of frustration. But then again, deep down he was somewhat relieved. He knew there was only one reason no one was telling him anything, it was something bad. Something else had happened when he was out – he was sure of that. And he was genuinely scared to find out what it was. But not all ignorance came with bliss. The anxiety was eating deep into his heart, mixed with guilt and unease. It felt like an ulcer eating away his walls; bringing pain and an unhealthy imagination with it. Bill folded his arms into fist, his nails biting into his palm. He wished that pain could relieve the one he was feeling within, but nothing worked. That was the point… nothing never worked.

Closing his eyes tight, he tried to remember what he saw or felt while he was blacked out to the world; but no amount of thinking could bring those missing memories back. There was one thing he remembered though – he remembered the comforting darkness that enveloped his whole body; he could liken it to the warmth of a mother's embrace. The natural feeling of love and tenderness that made him sleep longer. Opening his eyes, he was met with the white ceiling once more, and he could feel a smile tug at the corner of his lip; but it never came to fruition. Instead, a crease formed between his brows, and Bill could feel the crick in his neck due the uncomfortable position it was. As soon as he changed positions, the door to the room opened wide, and by the doorway was the doctor assigned to him.

"Good morning Bill, how do you feel today?"

"The same way I felt yesterday…" Bill replied shrugging.

"It would get better." The doctor assured, removing the stethoscope around his neck. Bill fought the urge to move or clear his throat as the doctor placed the end of the object on his chest. With a stiff expression, he managed to stay still long enough for the doctor to finish.

"You are getting better." The doctor beamed, wrapping the stethoscope around his neck once more.

"I still can't walk…" Bill began.

Sighing, the doctor moved closer, sitting by the edge of the bed. "I understand you might be frustrated, and of course you want to move along your normal activities, but first you need to go through physical therapy to be able to freely move around."

"Physical therapy?" Bill repeated, raising a brow up. "I thought you said nothing was wrong?"

"Physical therapy doesn't mean something bad. It is Uhm... it is a group of interventions designed to improve range of motion, strengthen muscles and joints, restore mobility and endurance and provide patients with medical-grade equipment and assistive devices. Bill you had a serious car injury, and before gaining full mobility, you need to work with physiotherapists to recover and learn to walk again. The journey from dealing with the immobile state that you're left in after a car crash to walking again can be arduous. However, if you work with a registered physiotherapist, that uphill journey will get you back to your pre-accident self and walking with confidence."

"Rehabilitation process..." Bill muttered.

"Yes, and due to the amount of time you spent in the hospital, it shows how severe your injuries were originally are. Once you are released from the hospital's care, you will need to visit a physiotherapy clinic. At the clinic you'll be paired with a team of healthcare professionals and physiotherapists who will be in charge of administering your rehabilitation program." The doctor explained, moving his hands as he spoke.

All Bill heard was the amount of time it would take for him to get better. All his brain registered was the fact that he couldn't walk properly until he got help. On his face was a dazed expression as he continued to listen to the doctor's words. "Your soft tissues will heal faster through movement." The doctor continued, "And mind you, you may be tempted to rest to avoid aggravating your injuries; however, this is not a good idea. The physical therapy program that is created for you will ensure that you get enough rest between exercises. We currently have a physiotherapist visiting the hospital; I will invite her over to explain so0me other things to you." The doctor said.

And true to his words, he brought her back; except this time, Bill wasn't alone with him. His dad, the physiotherapist and the doctor were all in the room with Bill. It didn't make him feel better; neither did it change the fact about what he had to do.

"Hello to you, I am Doctor Payne, But you can call me Susan." The woman greeted.

"Hello. I am Jordan Martins." His dad replied, shaking her outstretched hand.

"Doctor Greene told me he had explained some few things earlier on." She said, directing her words to Bill.

"Yes he did." Bill replied.

"At Floral worth Physiotherapy Clinic, we are dedicated to our patients; providing them with all the equipment, love and warmth they need." She began, and Bill was immediately reminded of a salesman here to sell a product. "You will be undergoing passive exercise first; since you cannot move your limbs by yourself, your therapist will move them for you. This allows for better circulation and keeps your limbs from atrophying." She explained. "And then you will undergo gait training. This involves showing you how to physically put one foot in front of the other. Following a car accident that makes mobility hard in your lower limbs, you will need to retrain your body on how to move. Your therapist will study your legs and hips to ensure your post-accident gait is strong and there is no chance of re-injuring yourself with every step."

Bill wished she could stop, he wished there was a better option that skipped all the steps; but there wasn't. He turned towards his dad, scanning his expression and realizing his father was already committed to seeing everything work out. "What comes next?" Jordan asked.

"Balance and coordination exercises. In order to stabilize you when you stand up for the first time after your car accident, you'll need to work on your balance. Using equipment like parallel bars, you will have your therapist stand behind you while you swing your legs in a walking motion. As for coordination, these exercises involve repetition of movement so that your body begins to remember how to take steps. Repetition is also a great exercise for people who have sustained a concussion after an accident, especially if memory loss occurs, because it helps the brain get back to normal. Then there is Neuromuscular reeducation, learning to use adaptive devices, muscle strengthening exercises, range of motion exercises, extra therapies, and then… dealing with trauma."

For every step she mentioned, Bill knew it was a long journey towards recovery. And that didn't sit well with him. "There's no way I can wait that long. I have to –" He paused, his lips pressing into a thin line. He knew he couldn't tell them the reason why he needed to leave the hospital as soon as possible. He wasn't sure if they were just normal workers, or if they were part of the government organization.

But with the doubts in his mind, he knew he couldn't proceed. "I need to leave." He said instead, looking at his father for help.

After you have undergone the aforementioned physical therapy, you'll move on to the last step in your recovery, which is walking. Ambulation exercises are designed to get you walking without assistance. You will need to re-learn how to climb up and down stairs, step over curbs and walk on uneven surfaces like pavement.

"How long does the journey to recovery take?" Jordan asked.

"Every patient responds to physiotherapy differently, so it's hard to pinpoint the exact length of each journey. The severity of injuries following a car accident can also determine how long you will need to work with a physiotherapist. At Floral worth Physiotherapy, our rehabilitation programs are tailored to the individual, and we also work to keep you from re-injuring yourself." She said.

Outside the hospital room, a man stood still, spying on the occupants of the room. As soon as the doctor turned towards his direction, he turned away, walking down the hallway like he had intended to do so from the very start. As soon as he was inside the elevator, he pulled his phone out, and dialed an unsaved number. "The target is awake." He confirmed.

CHAPTER THIRTY FIVE

John ran a hand through his hair for the thousandth time, yawning as he stared at the computer screen in front of him. His hair was now a mess, standing in different directions like a hurricane had ripped through it. His eyes felt gritty, and his neck had a crick that wouldn't go away. He had been awake all night, so was many others currently in the hall with him. Nibbling on his lower lip, he watched the screen fill up with a new string of codes, his pupils dilating as he watched it change.

"Any change?" A voice beside him asked, drawing his attention. It was Bell; the youngest tech advisor in the team they had assembled yesterday. She was just twenty one, and despite the age gap between her and the other members of the team, she held her ground and is on par with every other person. She reminded him of a younger version of Zendaya. She also had deep black hair that reminded him of the sky at night. Her hair was currently on the top of her head in a messy bun, but it did nothing but only accentuated the beauty of her nape.

"Any change?" She asked again, snapping John's attention towards her face.

"Ri – right…" He said, clearing his throat. "Yeah, Kind of." He added. Bell raised a brow up, a silent way of telling him to expatiate.

"After instantiating our session and manager classes, I created a new J Object to store our scan options in. I am trying see if the new commands would change, and perhaps we can find a pattern on how the other commands changed into auto pilot. These options directly correlate with the command line options we see from the Arachni tool when running Arachni –help since there's a lot." John explained, pointing it out on the screen. "And by storing a J Array with the values xss* and sql* in the checks option key, we tell Arachni to run XSS and SQL injection tests against the website, rather than simply spidering the application and finding all possible pages and forms. The audit option key just below that tells Arachni to audit links it finds and any HTML forms for checks we tell it to run. After setting up the scan options, we start the scan by calling the StartScan() method v and passing our test URL as the argument."

"So you will be using the ID returned by StartScan()?" She asked, tilting her head to the left as she stared at the screen.

"Pretty much." John replied, nodding.

"Then you should share your progress to the rest."

"The rest?" John repeated, looking around. "You don't mean…" He trailed off, horror on his face as he realized what she meant.

"Yep." Bell affirmed, stretching her hand out. "The whole team." She added.

"I … no way… I can't."

"It's okay." She replied, standing up from her seat. John watched as she lifted both of her hands up in the air, ironing out the creases in her spine as she stretched out, followed by a satisfactory moan. "That felt good." She muttered, adjusting the hem of her hoodie. Before John could predict what was happening, she was yelling at the top of her voice, startling him. "Everybody John just made new progress, and he would like to share it with us."

"What are you doing?" John whisper yelled, lowering himself in his seat. "There's no way I can address the whole team. I am... no way." he stuttered, groaning at the end of his words. But as fate would have it, he had no choice. Solemnly, John got up and headed towards the address podium. With each step he took, he could the cold piercing gaze of another human; the pressure on his shoulders continued to increase, until finally he was at the very front. At that moment, he wondered who would take a sixteen year old serious, but then again – it was a sixteen year old that flipped the entire world upside down.

John felt his breathing slow down as he stared at the crowd in front of him. Suddenly, the overhead lights were switched on, and one of the existing beams reflected directly on his face; blinding him momentarily until it was turned away. With the added lights, he could see their faces, and their expressions as they stared at him. "Uhm... Hi?" He laughed tersely, but none of the crowd was ready for humor. Clearing his throat once more, he hooked up the laptop to the projector; showing the whole room his progress so far. "The mobile app once consisted of a string of codes that act as a command for the operation of the Os, but by some strange twist of fate, something went wrong, and the command system is currently running on auto pilot. I created a new J Object to store our scan options in. and with the additional command system, I am trying see if the new commands would change, and perhaps we can find a pattern on how the other commands changed into auto pilot. These options directly correlate with the command line options we see from the Arachni tool when running Arachni –help since there's a lot." John explained, pointing it out on the screen. "And by storing a J Array with the values xss* and sql* in the checks option key, we tell Arachni to run XSS and SQL injection tests against the website, rather than simply spidering the application and finding all possible pages and forms." He explained, repeating his earlier words to Bell.

"What about the auditing system? We have tried that method before; but none of the commands passed the Os auditing system. It brought up the rear, and deleted the string of commands like it never happened." One of the teammates called out, stirring murmurs in the crowd.

"I faced that problem initially, but then I eliminated the concept data. The audit option key just below that tells Arachni to audit links it finds and any HTML forms for checks we tell it to run. After setting up the scan options, we start the scan by calling the StartScan() method v and passing our test URL as the argument. I retrieved the current scan status with GetScanStatus() w and then loop until the scan is finished, checking every second for a new scan status. Once this is finished, we print the JSON scan results to the screen x. The Arachni REST API is simple and easily accessible to most security engineers or hobbyists since it can be used with basic command line tools. It is also highly automatable using the most common C# libraries, and it should be an easy introduction for an SDLC or for general automatic use on your own websites for weekly or monthly scans."

As John continued to explain the process he took, his earlier nervousness dissipated, and his mind was filled with a field he thoroughly enjoyed. "For some extra fun, try running Arachni with your automation against previous web applications from the former codes with known vulnerabilities, such as BadStore. Now that we've looked at the Arachni API, we can discuss how to automate its RPC."

"I can help with that." A member of the crowd called out once more, drawing John's attention towards him.

"Can you please explain the concept to us. I am having trouble running past the RPC modulation." John explained.

"No problem." The man agreed, walking to the front. John pulled away, giving the centerpiece to the new host.

"The Arachni RPC protocol is a bit more advanced than the API, but it's also more powerful. Although also powered by MSGPACK, just like Metasploit's RPC, Arachni's protocol has a twist. The data is sometimes Gzip compressed and is only communicated over a regular TCP socket, not HTTP. This complexity has its benefits: the RPC is blazingly fast without the HTTP overhead, and it gives you more scanner management power than the API, including the abilities to spin scanners up and down at will and create distributed scanning clusters, thus allowing clusters of Arachni to balance scanning across multiple instances. Long story short, the RPC is very powerful, but expect more development focus and support for the REST API because it is more accessible to most developers." He explained. "Just one moment." He added, walking towards John's laptop. As soon as his hands touched the keys, he began to input additional information, changing the dynamics projected above.

public byte[] DecompressData(byte[] inData) { using (MemoryStream outMemoryStream = new MemoryStream()) { using (uZOutputStream outZStream = new ZOutputStream(outMemoryStream)) { outZStream.Write(inData, 0, inData.Length); return outMemoryStream.ToArray(); } } }
private byte[] vReadMessage(SslStream sslStream) { byte[] sizeBytes = new byte[4]; sslStream.Read(sizeBytes, 0, sizeBytes.Length);
if (BitConverter.IsLittleEndian) Array.Reverse(sizeBytes);
uint size = BitConverter.wToUInt32(sizeBytes, 0);
Automating Arachni 233
byte[] buffer = new byte[size]; sslStream.Read(buffer, 0, buffer.Length);
return buffer; }
private void xGetStream(string host, int port) { TcpClient client = new TcpClient(host, port);
_stream = new SslStream(client.GetStream(), false, new RemoteCertificateValidationCallback(yValidateServerCertificate), (sender, targetHost, localCertificates, remoteCertificate, acceptableIssuers) => null);
_stream.AuthenticateAsClient("arachni", null, SslProtocols.Tls, false); }
private bool ValidateServerCertificate(object sender, X509Certificate certificate, X509Chain chain, SslPolicyErrors sslPolicyErrors) { return true; }
public void zDispose() { if (this.IsInstanceStream && _stream != null) this.ExecuteCommand({"service.shutdown", new object[] { }, this.Token);
if (_stream != null) _stream.Dispose();
_stream = null; }

John stared at the projected screen, his lighting with each word it brought. This was in one way, a major breakthrough. Through this process, they were able to spot one of the patterns that pointed to0 the auto pilot. Although it was just one out of the many – it was enough.

"The constructor accepts three arguments u. The first two—the host to connect to and the port on the host—are required. The third one, which is optional v (with a default value of false), allows the programmer to automatically create a new scanning instance and connect to it, instead of having to create the new instance manually via the dispatcher. After assigning the Host and Port properties the values of the first two arguments passed to the constructor, respectively, we connect to the dispatcher using GetStream() w. If a true value is passed in as the third argument, instantiate Instance which is false by default, we create a unique name for the instance we want to dispatch using a new Guid x and then run the dispatcher. Dispatch y RPC command to create a new scanner instance that returns a new port, and potentially new host if you have a cluster of scanner instances."

John knew what he had to do, and it didn't take long before he took the reins once more. "With the new RPC, we can split the scanner instance URL for example, 127.0.0.1:7331into the IP address and the port 127.0.01 and 7331, respectively. Once we have the instance host and port we will use to drive the actual scan, we assign the values to our Instance Host and Instance Port properties, respectively. We also save the authentication token v returned by the dispatcher so we can make authenticated RPC calls later on the scanner instance. This authentication token is automatically generated by the Arachni RPC when we dispatch a new instance so that only we can use the new scanner with the token."

"Do we connect it to the second stream?" A voice asked, and John knew it was Bell.

"We connect to the scanner instance using GetStream() w, which provides direct access to the scanning instance. If the connection is successful and the scanning instance is alive x, we assign the Is Instance Stream property to true so that we know whether we are driving a dispatcher or a scanning instance (which determines the RPC calls we can make to Arachni, such as creating a scanner or performing a scan) later when we implement the Arachni RPC Manager class. After the constructor, we define the properties y for the session class, all of which are used in the constructor." He answered. New series of other questions came in, and each time – John was able to answer John held the entire room at attention, and it was a new kind of experience for him.

The Supporting Methods for Execute Command() Before we implement Execute Command(), we need to implement the supporting methods for Execute Command(). We're almost there! Listing 12-12 shows the methods we need in order to finish up the Arachni RPC Session class." He explained, pacing the room with each word he spoke. He could feel their gazes on him, he could feel the admiration with each discovery he made, and at that moment John felt the power he had in his hands.

"Listing 12-12 shows the supporting methods for the ArachniRPCSession class Most of the support methods for the RPC session class are relatively simple. The Decompress Data() method creates a new output stream from the zlib library available in NuGet, called ZOutputStream u. This returns the decompressed data as a byte array. In the ReadMessage() method v, we read the first 4 bytes from the stream and then convert the bytes into a 32-bit unsigned integer w that represents the length of the rest of the data. Once we have the length, we read the rest of the data from the stream and return the data as a byte array. The GetStream() method x is also very similar to the code we used to create a network stream in the OpenVAS library. We create a new TcpClient and wrap the stream in an SslStream. We use the ValidateServerCertificate() method y to trust all SSL certificates by returning true all the time. This allows us to connect to the RPC instances with self-signed certificates.

"We start by creating a public class called SqlmapSession u that will implement the IDisposable interface. This lets us use the SqlmapSession with a using statement, allowing us to write cleaner code with variables managed through garbage collection. We also declare two private fields, a host and a port, which we will use when making our HTTP requests. We assign the _host variable a value of string.Empty by default. This is a feature of C# that allows you to assign an empty string to a variable without actually instantiating a string object, resulting in a slight performance boost (but for now, it's just to assign a default value). We assign the _port variable the port that sqlmap listens on, which is 8775, the default. After declaring the private fields, we create a constructor that accepts two arguments v: the host and the port. We assign the private fields the values that are passed as the parameters to the constructor so we can connect to the correct API host and port. We also declare two stub methods for executing GET and POST requests that return string.Empty for the time being. We'll define these methods next. The ExecuteGet() method w only requires a URL as input. The ExecutePost() method x requires a URL and the data to be posted. Finally, we write the Dispose() method y, which is required when implementing the IDisposable interface. Within this method, we clean up our private fields by assigning them a value of null."

Someone at the very back left the conference room at that moment, and neither John nor the rest noticed who it was.

CHAPTER THIRTY SIX

He couldn't sleep; not a single wink. And for the first time in ten years, Key was awake all night. Seconds ticked by loudly, but he didn't move from that position. Key wished he could sleep, not think of anything, and just sink into the soft, deep embrace of his pillow and blanket; but life was not fair. He wished to stay that way for much longer. But reality was a different thing. The inner walls of his tummy clenched tightly, crying out for the supplement it was used to receiving; but Key had no intention of standing up, and going out. He had skipped both lunch and dinner yesterday, and the result of his actions was finally catching up to him.

The numbness of hunger was a strange thing. It left you tired, weak and didn't care if you were sad and happy. It demanded, and that was all it knew. Key was hungry, and the inner walls of his stomach clenched and unclenched as if waiting for something to come. The gnawing feeling started soon after, and the sides of his stomach began to bite in, leaving a rumbling sound each time it cried out for help.

"What am I doing to myself?" he muttered. "Such a fool…" Starving himself wasn't going to do anything, the world was still ending whether he liked it or not.

Standing up reluctantly from his bed, the coldness of the tile meets his bare feet causing him to wince as he rushed to the bathroom.

His eyes felt gritty, and as he came in contact with the bright lights of the bathroom he couldn't help but wince once more. Moving towards the sink, he grabbed his toothbrush and was about to add toothpaste when his eyes connected with the mirror above. The reflection staring back at him made me startled, causing him to chuckle as he scanned his disheveled appearance. By the side of his face was a depression made by the pillow, and with his eyes swollen and dimmed, and his face oily, he looked like a train ran him over.

"I look like shit..." he muttered to himself, rubbing the side of his face which was imprinted. He couldn't stop laughing as he realized the absurdity of the situation; one second he had been over thinking, and now he was laughing hard. The feeling of amusement subsided, and he was left with a feeling of nothingness, his eyes reflecting memories that he wished could disappear. As he rinsed his mouth, he came to a temporary conclusion. Today, he was going to do whatever he wanted. Closing the door of the bathroom slowly, he walked back towards the bed, a smile appearing on his face as he watched his girlfriend; who was peacefully sleeping. Almost as if she felt his stare, her eyelids twitched until finally she opened her eyes. "What are you doing?" She croaked.

"Good morning." He whispered, pressing a kiss on her lips.

"Hmm..." She mumbled, rubbing at her eyes. "Are you going somewhere?" She asked, her voice thick with sleep.

"No." Key responded, pressing another kiss on the side of her face.

"No?" She repeated, looking at him over her shoulder. "No work today?" She asked, absently scratching her chin lightly.

"I called in sick." He smiled.

"Sick…. Are you sick?" She asked, her face scrunching up as she pushed herself up.

"No."

She paused, hoping for him to say more.

"I took a day off to spend the entire day with you." He said.

His girlfriend stared at him for a few minutes, wondering if he was joking or being serious. "To spend the day with me?" She repeated.

"Yes." Key chuckled, finding her confusion cute.

"Nah, you are messing with me." She scoffed, laying back down and pulling the blankets over her head. "I am going back to sleep." she mumbled.

Shaking his head with an amused expression, Key turned towards the opposite direction. *'He might as well cook breakfast since he couldn't sleep.'* He thought, heading towards the bedroom door.

"You know you are behaving unlike you…" Jane said, resting her chin on her right hand.

"How?" Key asked, dropping a plate in front of her.

"You cooked breakfast, ordered my favorite for lunch, and now cooking dinner. You also did laundry."

"But I do all that on other days."

"Not all three…" She trailed off, slanting her eyes as she scanned his expression. "Since you came back yesterday you have been behaving weirdly. Is there something you want to tell me?" She asked. "Did you lose money again?"

Chuckling, Key dropped the utensils beside the plate. "No, it was just a bad day at work." He replied, shrugging. "Plus I was tired."

"Was it that bad?" She asked, reaching out for her wine glass.

"It was, but now it is resolved." He smiled, hoping she wouldn't ask any further.

As the night slowly passed away, they ate, they drank, and they talked. Every moment he spent with her was more special than it always was. Knowing they wouldn't do this after a time, made him soak in every bit of affection. He wanted to remember everything for as long as he could. "I am happy." Jane giggled. "Oh my, I think I drank too much wine." She giggled again, dropping the glass on the table.

CHAPTER THIRTY SEVEN

The unsuspecting member walked with a hurried pace, his nervous breathe filling his own ears as he kept glancing back. Turning around for the last time, he rounded the corner, entering the last room by the left. As soon as the door closed with a thud, he locked it behind him, taking a deep breath to expel the nervous tingles down his belly. Pulling out his phone, he dialed the number he already knew by heart.

Tapping his left leg, the dialing buzz filled his ears, his eyes darting from side to side as he waited for the person on the other line to pick.

"*Yes*?" A voice finally came through.

"New development." He reported, pacing as he filled the person the new details he just learnt. "Subject X is now devising a new tactic to regulate the patterns of the app. It's a breakthrough, and other members are in on it." He said.

"*Since when?*"

"Since now."

"*OK.*"

"When can I speak to the boss?"

"That's none of your concern." The person at the other end replied, his tone a mix of annoyance and impatience. As soon as the person said those words, the line went blank leaving the informant flustered.

"Stupid son of a bitch." He cussed, tucking his phone back into his. Clearing his throat, he smoothened the side of his hair, calming his nerves as best as he could. Breathing in deeply one last time, he opened the door, only to run into…

The man in the shadows looked up from the papers on his table, a crease forming between his brows as he heard the knock once again. "Come in." He muttered, and soon enough his right hand man walked in. "What's wrong? He asked, scratching the side of his beard. On his fingers were three rings, each symbolizing different meanings. The first one depicted a lion's head, the second an Infinity sign, while the last was a simple wedding band.

"New information."

"Let's hear it."

"Just got a call from-" Another knock on the door came in at that moment, cutting short his words.

"Come in." The man in the shadows replied, and soon the office was filled with one more presence; the third in command.

“New information.” He said, looking from the boss to the second in command, and then back to the boss. “Target O is alive. Woke up at the crack of dawn yesterday.” He reported.

“Yesterday?” The boss repeated, drumming his left fingers at the rim of his table. “Why is the report just coming in now? That’s twenty four hours too long. Has the target been moved?” He queried.

“No sir. The reason why the news is coming in late is because of the maximum security added to the facility. Plus, our inside man was stationed in a different place.”

“Too many excuses Edwardo, You know time is of essence.” The boss clicked his tongue, the crease between his brows deepening. “What about you?” He asked turning towards his right hand man.

“New breakthrough in stopping the app.” He replied. “Got the information few minutes ago.”

Silence ensued in the room, and yet the words were loud enough. They could all hear the sound of the wheels turning in their boss head. “Nothing seems to be going his way today.” The boss muttered, finally breaking the silence. “Call the informant on target X; ask if he can give us a copy of the whole operations. And better yet if he can't, let no one have access to him. If he needs another man, let me know. I will be trusting you with this Gordon.”

“Okay sir.” The right hand replied, earning a scowl from Edwardo.

"As for target O…" The boss trailed off once again, drumming his fingertips on the table in a rhythmic pattern. "Call in Jango." He finally added.

"Boss…" Gordon trailed off, his features etched with worry. Jango was someone they rarely saw. But when they did, catastrophe lay in his wake. Gordon wasn't a good man. He was a man who did what he had to do, when he had to do it; clinically was his style. But Jango's cruelty made Gordon seem like a good boy. The thought of him sent shivers down Gordon's spine. The man was a psychopath that had no regards for anyone, neither did he have empathy. In simple, he looked down on all lives forms, treating them like insects he could squash at any moment. The only exception was the boss. They had a history no one could understand. At times, it looked like they were enemies, and at other times they looked to be friends. It was a complicated thought whenever they were mentioned in the same sentence. "Alright." Gordon grudgingly accepted, swallowing down his protests.

"We have to do what we have to do. No mistakes this time." The boss replied, turning his back towards them.

Nodding his head in respect, he turned away, closing the door gently behind him. But almost immediately, the door was swinging open once more and Edwardo came out with a scowl.

"It must feel good to get complimented." He sneered, referring to their interaction inside the office.

"Not now Ed." Gordon sighed, walking away. But as always, Ed was on his heels.

“One mention of Jango and you suddenly have your tail tucked behind your legs.”

Gordon stopped immediately, startling Ed. He turned to look at him, his brows raised in disbelief. “You have no idea do you?” He muttered.

“What are you talking about?” Ed scoffed. “The man is no different, a loyal dog to the boss. Why do I need to know about him.”

Gordon shook his head, moving once more. Sometimes he wondered how naïve Edwardo could get. It didn’t help but he was the Boss's relative. He was a wildcard, he always kept his eyes on. “Stay away from Jango.” He said, “I hope you remember that.”

“Hey, you can’t tell me-" Before Ed could complete his words, Gordon walked off. He hoped Ed would listen this time. He really hoped so.

CHAPTER THIRTY EIGHT

Bill turned his face away; a ditch effort to appear uninterested. "Would you like something else?" His father asked, adjusting the blanket over his legs.

"Hmm hmm." He mumbled, still looking away. He was currently frustrated, and angry. He didn't know why, or who he was angry at… he just felt that way. He was also trying as hard as he could not to take his anger out on his father. He didn't deserve it after all; he was only trying to be helpful, and he could see that the days had taken its toll on his father.

"Do you want-"

"I am fine." Bill said, his words sounding short. Wincing inwardly, he looked down on his clasped hands. "I am fine." He repeated, but this time in a softer tone. "Sorry." He mumbled, unable to look at Jordan. He felt guilt, and he couldn't bear to see the hurt on his old man's face.

"Well… if you need me, you can just call. I'll be right there." His father assured, hoping for a reply, but he was met with silence. Finally giving up, he sighed and turned away. Bill's face remained frozen up until the moment he heard the door shut behind him. Glancing towards the door, he let out a deep sigh; the corner of his eye brows drooping low. Since the moment he was declared immobile, he had felt stuffy on his chest. A suffocating feeling that increased as the minutes ticked by. It felt like the walls of the room were closing in, and there was nowhere to escape. Wiggling his toes, he wondered how long he could do this. How long he could stay in this white world, and pretend nothing was going on outside the room.

For how long...

He needed to speak with his dad. And almost like he had been summoned, the door opened, and the room was filled with another presence. “Dad I-" Bill trailed off, a deeper crease forming between his brows. “Sorry, I thought it was his dad.” He apologized to the doctor that stood by the doorway. He was someone he had never seen, but then again – perhaps he had seen him, but wasn’t aware. He had been out for a couple of days after all.

“It’s okay.” The man replied, closing the door behind him. He was dressed in a doctor’s coat, and hanging from his ears was a nose mask covering half his face. “How are you doing today?” He asked, walking closer to Bill's bed.

“Same as yesterday… But, you’re not his attending doctor?” Bill queried.

“No, I am not. I am substituting for your doctor. He is currently in surgery.” The man replied, checking the Iv Drip attached to Bill's arm. “Has the nurse been here earlier?”

“Nurse Casey? No… she hasn’t stopped by yet.”

“Okay then, I guess there’s no issues with this.” The man replied, pulling out a syringe from his white coat. “This should help stimulate your nerves.” He said, reaching for the line.

“What do you mean?” Bill asked, suspicion settling in his mind. He expected the normal routine the doctors did by checking his vitals. But the first thing the man in his room did was to check the IV line. There was just something that didn’t seem quite right. “I don’t want it.” Bill grunted, yanking the line from the back of his hand. The force sent the contents of the IV splattering across, with a bit of blow that had back flowed.

“You need to calm down. You are only going to hurt yourself.”

The tension thickened in the room, and Bill came to a realization at that moment. This man was not a doctor. He was an intruder that was possibly here to kill him. “Who are you? Show your face!” Bill yelled, backing up. Although he didn’t want to admit it, he was scared and extremely vulnerable. He was in a situation where he couldn’t move without help, and here he was – facing a man that could possibly harm him. “Dad!” He shouted, his voice sounding hoarse as he yelled across the room. But the intruder wasn’t taking any chances. He lunged across, tackling Bill on the bed. With the strength he could muster, Bill tried pushing the man away, grunting as the weight made his arm tremble. “Get the fuck away!” He cussed, hitting the man’s face with his elbow. The struggle continued for a few more seconds, before finally the door burst open.

In shock and horror was a father who saw a stranger attacking his son. His first instinct was to protect. He had no care for his own life. All Jordan wanted to do at that moment was pry the man away from Bill. As he ran across the room, the intruder looked up at that moment, and Jordan could see a moment of panic cut across his eyes. "You son of a..." He cussed, throwing a punch on his face. Almost immediately, something clattered to the ground but Jordan wasn't paying attention to what it was. His attention was solely on causing as much damage as he could to the intruder.

But the stranger had other plans, he wasn't going to take this lying down. Ducking, he threw a punch to Jordan's midriff, knocking the air out of his lungs. But Jordan couldn't wait, he pulled on the coat he was wearing, twisting it around so he could lock one of his arms. Struggling to be free, the intruder tried twisting away, but Jordan couldn't afford to let go. The intruder walked backwards in a fast pace, pulling along Jordan who was imbalanced for a split second. The crack in offense was all he needed. Taking the chance, he yanked his arm, the sound of tearing fabric filling the room. Throwing the piece of cloth away, Jordan tried to attack once more, but he had no leverage. He was now on the defensive; trying hard to block the blows that were aimed for his vital points. Grabbing a small plant from the floating shelf, Jordan threw it at him; hitting him square in the chest. Although it looked inconsequential, the ceramic pot had caused some damage. Right next to the intruder's feet was soil, and the jagged pieces of the ceramic pot.

With the external help, Jordan was able to keep some distance between them both. The tension in the room was unreal, and the sound of rasped breath, and sweaty skins mingled to create a blockbuster moment. "Who sent you?" He queried. "What do you want? How did you get here?" Jordan asked, but he was met with silence. He knew there was no way the intruder was going to answer his questions, but that didn't stop him. He was very frustrated, and for now he was sure the government was not going to keep his family safe. It was all on him, and he had a responsibility to carry out.

Reaching down, the intruder wrapped his hand around one of the bigger pieces; holding his hands out in a punching stance. Another battle ensued, but this time Jordan was in the losing end. The sound of painful grunts, and smacking fist echoed and bounced around the walls.

“Dad watch out!” Bill yelled.

Blocking another jab towards his midriff, Jordan punched back harder; causing the intruder to stagger backwards.

The intruder looked at his watch, then back at Jordan. And in the most bizarre of situations, he did the unthinkable. He turned around and ran straight for the glass windows. Jordan thought the man had gone crazy, but he had no thoughts of stopping. The situation suddenly flipped over and the man jumped through the glass windows, leaves shattering sound behind as he fell to his death. Jordan felt a piercing sound fill his eardrums, blocking out all the other sounds as he stared at the gaping hole in the window. Subconsciously, he walked forward, slowing his pace as he stood by the edge of the windows. With the wind wiping against his ears, he could finally hear everything again; including the sound of horror carried by the wind. The sound of screams and sirens bellowed from below, causing him to swear once more.

“Dad…” Bill called out, diverting his attention towards him for the second time. His sons face was as pale as a ghost. Sweat had broken out on his temples, and Jordan realized once more that his son had come close to dying. With no words said between, he circled his arms around Bill's frame, hugging him. “You are fine buddy. You’re going to be fine.” Jordan assured, hugging his son tighter.

“That’s not… I don’t think so.” Bill muttered, his breathing sounding labored. Pulling away, Jordan looked at his son’s face clearly, and knew he was trying to say something.

“I don’t know…” Bill mumbled, holding on to his upper arm. “He injected me with something.” Bill confessed.

“What…” Jordan trailed off, his kind traveling back to a few minutes ago when something had dropped to the ground. “Fuck.” He cussed. The door burst open, and the guards were finally present.

CHAPTER THIRTY NINE

The wind from the south blew in gently, caressing Key's face and ruffling his hair as it passed him by. It was times like this that he could pretend everything was okay. He closed his eyes towards the direction of the wind. Raising his left hand, he lightly traced the scar by the side of his neck, and he was temporarily transported back into the past – to the day he almost died. He could remember the suffocating fear that embraced his whole body as he realized he was going to die on for a country he sacrificed his whole life for. His unheard screams, the sound of strangers yelling… the scent of blood, and the ticking sound of his death as it approached closer filled his ears. It was a day he would never forget, for as long as he could live. He had lost faith in humanity at that moment, and he realized that there was no turning back once he died. But then again, life had a way of playing jokes on the ones who didn't see the full picture. He remembered the one who had helped him; the one who had extended hope at his darkest hours.

"Key." A familiar voice called out to him from behind, breaking into his thoughts, and clearing his mind. Even though he didn't turn around, he could tell who it was. He remained silent as the footsteps drew nearer, and continued to remain that way until the person stood beside him. It was his girlfriend.

"What are you doing?" She asked, and for the first time since she appeared, Key looked at her. From her curly brown hair, to her green eyes, and freckled cheeks; she looked absolutely gorgeous to him. He couldn't help but feel pride swell in his chest as he stared longer.

"Is something on my face?" She asked, touching her left cheeks in a conscious manner.

"There is."

"What?" She asked, patting the other cheek.

"Beauty."

She paused at that moment, her mouth wide open as she stared at him. "You say the corniest lines sometimes." She grumbled, eliciting a chuckle from Key.

"You've been outside for a while. Are you doing something?"

"Nothing." He nodded, dropping his left hand back to his side. "Just enjoying the breeze. Did you miss me?" he asked.

"Well…" She trailed off, moving closer to where he stood. "Maybe." She added, pressing a lip against the grin on his lips.

"Okay." Key grinned, wrapping his arms around his waist. He buried his face on her hair breathing in a familiar scent that calmed his nerves.

"What's wrong?" She asked, her voice sounding muffled against his chest.

"No… I mean, yeah, no." Key replied, folding his arms against her back. She could tell he was nervous, but she wasn't sure why he was nervous.

"Should I quit my job?" Key joked, an underlying question of what he truly felt.

"Quit?" She repeated, raising a brow. "Why do you want to quit your Job?"

"I was just thinking." He answered, rubbing the side of his neck.

"Well…" She paused, biting her lower lip as she contemplated something. "I know you love your job. But if, you really want to quit, you don't have to worry. I will take care of everything."

Key felt emotion clog his throat, while he held on tighter. "I am not quitting." He mumbled, resting his chin against her hair. "But thank you." He added, pressing a list in the middle of her head. They stood that way for a minute, before finally – Key let go.

"We should go back now." He said, forcing a smile as he pointed towards the direction she came from. There was definitely something he wasn't telling her. Pausing, she raised a brow, scanning his expression once again. She wanted to ask so many questions, but then again she knew the nature of his job. She knew every information fell under confidentiality and secrecy. "You're not getting hurt, right?" she queried, her stance showing she won't budge if he didn't reply honestly.

"I don't know yet." He confessed.

Catching the seriousness in his voice, she held on to his arm almost immediately. “What happened?” She asked.

“We need to go back home, there is something I need to do.”

“Is it urgent?”

“Very… but…” Hesitating, he put her hand in his. “Do you trust me?”

“Yes.” She replied with no hesitation.

They had to go back home.

The sound of anguish and pain reverberated across; and it continued to echo down the hallway. The closer someone got, the louder the cries for help became clearer. At the very end was a room with the door slightly ajar.

“You stupid fool!” The boss yelled, slamming the edge of the baseball bat against Edward’s head. “How dare you go behind my back! Look at the fucking mess you caused!” He yelled, slamming the bat down once more.

Gordon watched from a distance; an unreadable expression on his face. He should have known Ed was going to do something. His stupidity was what was costing him right now. There was blood everywhere. At the bottom of the baseball bat, the rug, the boss pants', Ed's shirt. A loud thwack followed once more, before finally the boss dropped the bat with a thud. “Anyone by the hallway?” He called out, directing his words to the guards standby. “Take this useless son of a bitch out of here. I don’t want to see his face.” He cussed, spitting few inches away from where Ed laid in the pool of his own blood. In sync, two of the guards rushed in, dragging out Ed with a trail of blood left behind.

“Fucking Imbecile.” The boss cussed again, heading towards his table. With hands that tremble with fury, he picked up a cigar, clipping the end, and lighting it up. As soon as he took a long one, the tension radiating in his shoulders let go, and his nerves were visibly calm all along, Gordon didn’t say a word. He watched on with the same expression he had all along.

“Who died?”

“Sebastian amoy.”

“One of the suicide forces?”

“Yes sir.”

“Hmm…” The boss mumbled, sitting down with enough force to shift the chair backwards. “It’s a pity we lost one of those… I raised them for one purpose… and that fool…” He trailed off, drawing another length of cigar. “Take care of it like we always do. Compensate his family.” The boss said, placing his left hand on the table top. Unlike his blood stained pants, his hands were surprisingly clean; and had no blemish on them.

“Yes sir.” Gordon replied, turning around to leave.

“Wait a minute.” The boss called out, stopping him in his tracks.

“Sir?”

“Jango?”

“Contacted.”

“When can he get here?”

“Two days from now.” Gordon replied.

“That’s too far.” The boss grumbled, running a hand through his hair. “Tell him he has to be here tomorrow. Time is of essence, and with that fool messing things up; I need the plan to proceed smoothly. Between, release something… something that can sway public's opinion.”

Gordon felt his shoulders tighten at the mention of the name. He wished he didn’t have to call again; but then he couldn’t risk the boss wrath either. “Yes boss.” He nodded, finally exiting.

CHAPTER FORTY

It is always hard to go back. Always hard to pretend time has not passed by for the things we once abandoned. Although, Key had new resolve, the determination to release to the world the biggest conspiracy ever needed a much bigger courage than he thought. He had to tread carefully, for the people he was dealing with were monsters. The moment he had joined the summit, that was the moment he had unofficially signed a confidentiality agreement. But the secret decided the fate of millions of people he held dear; including his family and closest friend, there was no way he could pretend to remain oblivious. He had tried that, it didn't work. Breathing deeply, he strode towards the place he once proudly called second home, an unreadable expression on his face as the elevator's door opened to the floor his office was.

As the doors opened, the first thing Key saw was a woman dressed in red all through. The red reminded him of blood, and almost like she heard his thoughts she smiled with lips painted in the same color. "Pretty, isn't it?" She asked.

"Pardon?"

"The color… isn't it pretty?" She asked.

"I guess..." Key replied, feeling strange at her question. As he slightly turned to move the other way, she stretched her hand out, impeding his movement. "Nice to meet you, Mr....?"

"Detective Key." He introduced, reaching forward to shake the outstretched hand.

"Have you seen the news lately?" She asked; her voice grating on his nerves the longer he listened. Not only was it weird, his instincts was telling him to stay far away from her; she reeked of death.

"Yes, I watch the news every day." He smiled, nodding to the opposite direction before he tried moving once again

"Well, you might need to watch Today's special. I promise it will be interesting. There is a lot to learn." She said suggestively, and just like she appeared, she was gone. Key couldn't help the frown that was left on his face. 'Who was she, and what was she talking about?' he wondered. All in, it was a strange experience.

The sight of the busy floor filled his sight as soon as he stepped beyond the doors. '*What was the point of working so hard?*' he wondered as he stared at everyone running and working like their life depended on it. At exactly three O clocks, he spotted his direct subordinate – Ellie. Signaling to her, she nodded her head and weaved her way through the crowd, walking towards him. "Sir you are back, you came back earlier." She greeted.

"I guess I miss work." He guffawed.

“Miss Work… well, you came at the right time.” She smiled tiredly, indirectly pointing out the busy atmosphere.

“Did something happen again?” He asked. Although things got so busy around here, there was something about the atmosphere that made you get dizzy just by watching.

“Yes. We just found out the force that has been manipulating and expanding the stock market; leading to the near economic crash that happened a week ago.”

“What?” Key repeated, not sure he had heard right. “Who was it?” He asked, feeling a lump settle in his throat as he waited for the answer.

“Eco Concept. Uhm, they are a group that supposedly deals with Eco ideals. It was only a front. A paper company.” She clicked her tongue.

Key wanted to scream that they were all wrong. He wanted to yell that it was all a false flag and the real culprits was a pyramid scheme, and the fact that the government had joined in on the game. They were planning to abandon them all. To build the ship, lots of money was needed, and it was no surprise they were using this way. But then a whole conglomerate group was turned to a paper company overnight. “Place the report on my table, I will be back.” First of all, he needed to catch the one person that raised the biggest flag. “I will be back…” he said, turning towards the opposite direction in an almost stiff like manner.

As the elevator rolled down, Key felt a new wave of nervous fever rip through his body, and his stomach once again – began to hurt. He didn't realize he clenched his teeth so bad, and it wasn't until he breathed through his mouth that he realized it was beginning to hurt. Lowering the intensity on his jaw, Key look forth; and as the elevator door pinged and opened, he strode out, and walked directly towards the building entrance; scanning the perimeter for a woman that brought the biggest flag, a woman in red.

Frantically searching, Key sifted through the faces; but every face his eyes connected with was not the one he wanted to see. Stepping out of the building, he headed towards the parking lot; praying to God he didn't miss the chance to talk to her once again.

But this time, God didn't answer his prayer. She was nowhere to be found.

"When you explained earlier did you say the Post Request with Sqlmap method accepts four influences?"

"I did. I said the first influence is the URL that will be sent to sqlmap. The second influence is the data that will be in the post body of the HTTP request—be it POST parameters or SOAP XML. The third influence is the value that will be passed in the SOAP Action header in the HTTP request. The last influence is the unique value that is vulnerable. It will be replaced with an asterisk in the data from the second influence before being sent to sqlmap to fuzz. After we bypass the message to the screen we might be able to get a contemporary URL; and once you've tested it out, it will bring this-." John explained, showing her on the computer screen.

"Yay!" a loud holler filled the hall.

John looked up at the exact moment Bell did; their eyes drawn to the commotion that was garnering attention at the back. “What happened?” He asked, turning towards her.

“I… I don’t know.” She replied absently, standing up from her seat. “I will be back.” She added, before walking towards the back. John watched her leave, and he wondered if he should also do the same. Giving in to the curiosity that bite deep down in his belly, he finally stood up, heading to where they all stood. He paused as he stood by Bell's side, his brain trying to connecting the pieces of conversation they were currently having.

“Did something happen?” He asked once more, finally giving up since he couldn’t catch on.

“The financial crisis is over.” She stated. “

“Over?” John repeated, still not connecting to her words.

“They discovered what company was behind the recent financial breakdown. It’s a paper company from Russia. They got traced back to their base... some good news for once.” Bell mused, placing her hand akimbo.

“Do you believe that?”

“Well… I don’t know to be honest.”

John bit down on his lower lip thinking about the current news. It seemed sudden, and too easy considering what damage had been done. The puzzles didn’t seem to connect, and that left behind a jagged path.

"John." A voice from behind called out to him, causing him to spin around "Yeah?"

"You have a guest."

John was taken aback. A guest? The thought of someone coming to visit him had never crossed his mind. He was aware of the fact that, his current location was hidden from the public.

So who was it? A foe or a friend?

CHAPTER FORTY ONE

Over and over again; a vicious cycle that wouldn't let him go. Jordan felt numb. There was no better word that could describe what he felt at that moment. It was like life was telling him to go fuck himself. It was like life was playing a game on its own terms; terms that wanted his son dead at all cost.

"He will be fine." The doctor assured once more, patting Jordan's shoulders as he walked away. But Jordan wasn't fine. He wasn't fine at all. Fortunately, Bill had been injected with a strong sleeping tranquilizer. But just… *what if that wasn't the case? What if he had come late?* The different scenarios that ran wild in his head wouldn't let him go. Letting out the air in his lungs slowly, Jordan knew he had to make a tough decision. The government couldn't keep them safe. The signs were there all along… how could he miss that?

They claimed maximum security, yet – they got Bill each time. Were they lying, or was the opponent that strong?

"Dad…" A groggy voice called out to him, and it didn't take a genius to know Bill had finally woken up. His eyes were out of focus, his pupils pale as he tried to concentrate on whatever he was currently looking at. "Dad…" He called out again, distress etched in his features.

“I am here.” Jordan replied gently, holding on to Bill’s left hand. He lightly covered the top with his other hand, gently patting as if to calm his nerves.

“Are you alright?” Bill asked.

“Silly… I should be asking you instead.” Jordan said warmly, sitting by the tip of the bed. They sat in silence, the unspoken words raging louder than any other thing they’ve ever shared.

“I need to see John… I need to help him.” Bill started.

“John has all the help he needs right now, you don’t have to –”

“Dad…” Bill pleaded in a broken voice, his lips quivering as he stared at Jordan. “Please…” He repeated, shaking his head as he tried to fight back the tears that were threatening to fall. “They need me…”

Jordan tore his eyes away from his son’s face, a huge lump forming in his throat. With his back turned towards Bill, his hardened expression fell flat, and the look of angst and worry were etched on his features. What could a father do? What could he do?

Bill watched his father’s back in rigid silence; his clasped fingers clenching and unclenching as he waited for a verdict. The silence seemed to exist, until finally his father turned back towards him. His eyes were rimmed red, and his lips were twisted in a regretful smile. “They say sons grow up very fast. I guess they are right.” He joked.

Almost immediately, the suffocating air turned normal, and Bill could let go of the breath he was holding. Unclasping his fingers, he covered his face with both hands; hiding the look of relief that washed across his features.

In John's presence were three guests he had not expected to see. People he would consider acquaintances, and nothing more. "Hello…?" He greeted, hesitation tinged in his voice as he stared at them all.

"It's been a while." One of the men responded, and John found himself a bit confused.

"Detective…"

"Key." The man responded once more, filling in the gap John was missing.

"Can we go somewhere quiet?" Marcus added, motioning towards the entrance they had just come from.

"Yeah, sure… uhm…" John looked around; wondering where was the best place to hold the conversation. Thinking hard, he decided on a meeting room they rarely used. "I will be back." He said, directing his words to Bell as he left with them. Their footfalls echoed down the hallway as John led them into the conference room. "Do you want anything, coffee… juice?"

"No, I am good." Marcus replied, shaking his head.

"Same." Key replied, while Victor shook his head also.

“Well, alright…” John trailed off, feeling awkward. He wasn’t sure about what to say or do at the moment. But his curiosity was blossoming. He wanted to know why they were here.

“I think I might know a way….

“What’s the answer?” Key asked, biting his index finger as he waited for Marcus to answer.

CHAPTER FORTY TWO

The air in the room thickened within minutes, and the nervous pit-a-pat of John's heart felt louder than it actually was. No one said a word as they waited for Marcus to answer the mystery he had created.

"Come on… say something." Key groaned, folding his arms against his chest.

"Hold up. I am trying to deliver the answer in the best possible way." He protested, holding a hand up as he spoke. But his actions held the opposite effect. Almost immediately, groans cut across the group; each wishing they could knock him out.

"Okay. I have already gathered my thoughts. It is…" He paused for the umpteenth time, and just when another second passed by, he finally confessed. "Hypnosis. The answer is hypnosis."

"Hypnosis?" John repeated, a frown creasing his brows. Of all the [possible answers he could think of, Marcus answer was not one of it.

"Hypnosis?" Key also repeated, raising a brow up as he stared at Marcus.

"I can explain. This is not just any hypnosis, but a beta Alpha dex I created.

"Think about this. There is only one person that has survived a disappearance from the app. A living testament –"

"I don't even remember that time frame. I mean, it doesn't exist anymore." John protested.

"I know… but hear me out. What if… I know, hypothesis; but at this point, I can believe almost anything. What if those people that disappeared didn't vanish, but are currently living in an alternate universe we can't see. Perhaps we are just separated by a thin veil. Think about it for a second. The beta dex I created taps into the brain waves, and tries to replicate the memories it had once gone through."

"This sounds like some Sci-Fi shit."

"And you think the app disappearances aren't Sci-Fi already?" Marcus retorted, tilting his head to the left. Key threw his hands up in retreat, and everyone was left to ponder on the new piece of information. The one person that had so much to think about was John. It would be a lie if he said he constantly thought about supposed disappearance. But then again, how could he think of something that truly didn't exist. He was currently living in a fast paced life, and every moment was exhilarating in its own form. He could feel a tingling sensation pointed towards his head, and he realized that someone was staring intently at him. Once he looked up, his gaze caught Marcus' eyes, and he knew what that look meant. The decision was in his court. They were all waiting for him to make a stand. "Does it hurt?" He queried, trying to bide time for as long as he could.

"Not that I know of." Marcus confessed, rubbing his nape.

"What do you mean..." John trailed off, narrowing his eyes as he stared intently at Marcus. "Have you ever tested it out on someone else?"

"Well." Marcus answered, looking away. "I am about to." He muttered.

"Come on!" John groaned, throwing his hands up. "There's no way I am going to be your test animal."

"Don't worry. Trust me, the design is very human." Marcus assured, thumbs up. As John stared at Marcus lopsided grin, he wasn't convinced in any way; yet – ten minutes later he found himself seated in a chair that reminded him of the dentist seat, arched back so he was half sitting and half sleeping. On his very head was a metal band Marcus had sworn wasn't dangerous, although he had never tested it out. "Guys... I don't know..."

"Shuuush." Victor responded, trying his best to make John relax. Left with no choice, John turned towards Key hoping to find an alley; but instead he was met with a man that was trying his hardest to look sideways."

"One, two... go. The machine is connected." Marcus said excitedly, staring at the laptop in front of him. John frowned as he watched Marcus and Victor's happy reaction. He wasn't sure if they were happy about their invention, or the fact that it was working.

"How do you feel?" Victor asked, turning towards him.

"I don't know." John replied truthfully.

"It's alright." Victor assured once more, before turning towards the laptop. Soon enough, John ears was filled with a beeping sound; a sound that seemed to only exist to him. "Uhm… can you hear that?" He asked, hoping to confirm.

"Hear what?" Key asked, finally moving closer.

"The beep beeps sound…?" John asked, stiffly turning his head towards Key who was standing by his left side.

"I can't hear a thing." He confirmed. "Is there a problem?" He asked, directing his question to Marcus.

"Oh, no worries. It's just the brain waves replicating. Perfectly normal." He added, grinning once more. "You need to relax; your brain waves are going crazy right now." Marcus called out.

"I don't think I can be calm as long as I have this on my head." John muttered, closing his eyes. *'Never doing this again.'* John thought at that moment. Twenty minutes slowly passed by, and John was beginning to feel the angst that came with sitting in one position for too long. "How long do I have to stay this way?"

"Not for much longer." Marcus replied.

"Have you found anything yet?" Key muttered, placing a hand akimbo as he stared at the screen. He couldn't understand what was going on right now, but he presumed the blinking waves on the screen were John's brain waves.

“We don’t know yet. All we are doing right now is collecting as much wave length as we can. We would have to sort through it all after we are done.” Victor shrugged. John listened in, biting the tip of his tongue to stop himself from groaning.

Not much longer went on for another forty minutes, and John was about ready to jump out of the chair. He was mentally tired.

“I –”

“Done.” Marcus cut in; eliciting a scowl from John. As soon as the band was taken off; John felt like he had lost something. It felt like his head was empty, and that sent Goosebumps down his arm. Although the band wasn’t heavy, the weight it carried definitely made a difference. “I feel light hearted.” John groaned, running a hand through his hair.

“Take a breather.” Marcus urged, jutting his chin towards the entrance. “It might help.” He added.

John bit lightly on his lower lip, contemplating that option as he stared at the door. ‘Should he?’ he wondered. It wasn’t like anyone had barred him from going outside, but then again there was always the invisible fear that someone was always watching; ready to snatch him up if he ever let his guard down. “ I don’t know…” He muttered.

The silence continued for a few more minutes, before finally he gave up. Standing up in one go, he stretched his hands above his head, grunting as the sweet relief washed down his body. “I will take that offer.” He said sheepishly, rubbing the back of his nape. “I will be back.” He said, before finally heading outside.

There was something different about being embraced by the natural breeze. It slowly washed against your skin, teasing the hairs it cut across while cooling the temperature of your body. The moment was better than a thousand air cons. John was not someone that was sentimental. No he wasn't. But then again, witnessing a time period that could be end of the world left you feeling pretty much unsettled. Tipping his head backward, he stared up at the night sky, and the first thing he noticed was how still it was. It was like the world was at peace, and everything that was happening was only a dream. "How ironic." He muttered, his lips twisting into a wry smile.

But irony was far from over.

As soon as he muttered those words, headlights of a camouflaged van he had not seen switched on; and the light pierced through the darkness, blinding him temporarily. Squinting against the bright light, John wondered who it was. The play was just beginning, and the climax of the story was yet to be over. Automatically, the intensity of the headlights reduced, or perhaps it was because John's eyes had somewhat adjusted to the luminance. Whatever it was, John could see clearer. The van approached, and didn't stop moving until it was full inches away from where he stood. With enough accuracy to scare the soul out of his body, the driver of the van swerved right, startling John who moved backward. "What the fuck…" He cussed beneath his breathe, placing a hand against a chest as he tried to see who the terrible driver was. But that question remained unanswered as the door to the van swung open with enough force to make him flinch.

"Hey, what the fuck are you…" The words died on the tip of his tongue as he looked inside the van. John could feel the horror rip through his body, his eyes twitching, and his breath coming in short gasps. There were no words. Not a single letter uttered as he stared at the occupants of the car. There was no need to. He knew at that moment what he needed to do. Closing his eyes against the decision he had to make, he regretted not acting faster. But then again, what could he do? He scoffed. Opening his eyes, he nodded his head, and just as he took a step forward, he turned around one last time.

Life was Picasso. It painted coincidences like the master it was. And at that moment, it was painting a masterpiece. As soon as John turned around, his gaze caught the gaze of another. Although brief, it was enough. It was enough time to leave a vital piece of information.

CHAPTER FORTY THREE

The night sky has witnessed it all – love, war, murder… and currently the two cars that raced dangerously closed to each other. Tonight was a night packed with danger, the need for speed, and a chase for the truth. Two vehicles raced treacherously close along the highway, gaining and loosing distance as they continued to drive. But it wasn't as simple as it looked; it was a race filled with acceleration, the burning desire to win, with their lives on the line. The man in the car pressed down on the gas pedals; going two hundred miles per hour, the gas cylinders exploding in a ball of smoke, while his eyes smoldered as he chased.

Flashback

Key failed miserably as he tried to stifle a yawn that was pushing through his throat. Giving up midway, he let it go; relishing the feeling of absolute abandon. A sheepish grin cut across his face, making his appearance somewhat goofy to the third eye. Sniffing, he looked around the silent car park. But as fate would have it, his eyes moved towards the direction that was unsuspecting. He could see a van, and a figure that seemed too familiar to be ignored. With both hands in his pocket, he walked closer to where they stood; his curiosity increasing with each step he took. An important clue appeared at that moment, and Key could fit in the two uneven pieces together.

Something snapped in his mind right away, and he realized who it was all along. "John!" He yelled, picking up his pace as he ran after the van. The driver of the vehicle saw an unwelcomed figure running towards their direction, and did the best he could – he stepped on the gas pedal. Before Key could meet up with whoever it was, the van zoomed off – leaving behind dust, with no figure in sight. At that moment, Key knew he had to act fast. There were too many variables that were involved. One. Who were they, and where were they heading to. Two. Key had seen it all, he knew John had gotten into the van without any exchange; why? And, should he go back, report the entire situation to Marcus and Victor. Could he afford to lose more time and do whatever he pleased? Each and every question ploughed his mind, leaving him distraught for a split second. Swearing beneath his breath, he touched around his pockets, wondering if his car keys were there. The relief that went through his spine was short lived, since he knew it was only the beginning of a battle. Although it seemed like minutes had passed,

"Did you see that?" A voice from behind stopped him before he could move, causing him to twist his neck so fast he almost had a whiplash.

"Down to the last scene." Key replied, pressing they key to his car. His car is a Maverick, and although it isn't the flashiest vehicle in the ford lineup, it is a small truck that may redefine what a truck can be.

"I am coming with you." Bell declared, getting into the shotgun position.

"There's no way in hell kiddo."

"Hey, I am not a kid. And just so you know, I was trained specially in the ops, before I decided to work here."

"What are –"

"I will be fine." She cut off. Key swore beneath his breathe, contemplating what to do next. Tightening his fist, he swore once more, grudgingly driving after the van that had left few minutes ago.

Back to present

The wind howled as the cars cut across the wind, the tires burning every single time it connected with the road. The man behind the wheels was no other than Key Oakley; an ace in the new detective team, and a veteran on the road. The car Key was driving was a gift from the FBI, a power horse that made the chase easier. He was driving a Red Maverick; a car which can reach the speed of two-sixty miles per hour, and go from zero to sixty in two point nine seconds. The Maverick pumps out over nine hundred horse power, and its carbon fiber body makes it lightweight, aerodynamic and very pretty to look at. Key handled the beast quite well, the streets lights becoming a blur as the car whipped across the road.

"How the hell are they still so fast…?" Bell cussed while trying to steady her hands so she could shoot the target in front. "Are you serious about being trained?"

"Yes." Bell replied, gritting her teeth as she held on for her dear life. Thinking for a split second, Key reached for the hidden space between both seats and pulled out a personal gun that he had bought three years ago. The gun was capable to push a three hundred grain bullet out of the barrel of its Magnum Research Desert eagle at 1475 fps and generate almost as much muzzle energy as the .45/70. Key stared at the gun thoughtfully, praying he wouldn't regret this; but his hands were quite full.

"You want to take over the wheels… switch positions?" Key asked over the loud purr of the engines.

"Switch positions?" Bell repeated, looking at Key like he had gone crazy. "How the hell do you think that will work? Do you know how much mile per second you going right now? We will die before we switch over!" A different car came from nowhere, swerving right in front of them; causing the car to jerk forward as Key applies the break quickly. Key wasn't the only person to suspect the new addition; Bell's eyes were narrowed at it. Proving them right, a bullet from the new car in front hits the windshield, heightening their suspicions. Whoever was in the van John had entered willingly was bad news.

Bell let out a string of cusses, returning back to her position. But Key doesn't flinch; a ruthless determination driving him on despite the velocity at which he was flooring the car. His heartbeat didn't quicken because of that; but it quickened because of the excitement that filled his bloodstream with each second that passed as the chase continued. "I guess you're glad I am here." She mused, collecting the gun from him. Before Key could reply, Bell lays low in her seat as she began to shoot. The sound of dropping shells echoed in the confines of the car, until finally one of the bullet connects; blowing off the tire of the car that the earlier shot had come from.

"Hell yeah." Bell whooped, excitement apparent on her face as she continues to shoot.

Another shot rang out from behind, and Key knew it came from another car that was planning to take them down. It was a perilous fight. The fight for dominance continued, the bullets ricocheting off the car's exterior as neither of them was letting down.

The sound of loud blades whipping against the wind caught Key's attention, causing him to look out the driver's window. "It is either a foe, or perhaps the unit realizing John is missing" He commented. The homeland helicopter was now hovering from above, casting a strong beam of light upon the muscle cars, Key and the Bell were chasing. "They are friends." Bell muttered, adjusting her hold on the gun.

"I am thankful for that." Key replied, tightening his hands around the steering wheel. "They're damn persistent." Key groaned, observing the cars in front. With the help of the helicopter's light, he could see that the cars in front were riddled with bullet holes, with some of the glasses broken, but still the car was still somewhat alive, pushing on and managing to evade the high speed Key was flooring the car. He wondered what model their cars were, and if they were relatively higher than the Maverick he was driving. The chase had now gotten off the highway, and was now inland streets; which meant people and other cars were involved now. The two suspecting cars in front took advantage of that fact, their cars screeching as they turn around, and drove into a street corner block. The helicopter and Key's car follow, alongside more gunshots.

"Shit… I hate this." Bell cussed, knowing she had to be extra careful when shooting now. There was no way Bell could afford to hit a normal civilian; though the assholes in front had no problem of doing that since they continued to shoot regardless of where they were.

Key realized the helicopter's beam was now pointed at them, and he wondered if they were trying to pass across a message. Putting on his truck's radio, he turned the frequency, hoping he would be able to connect. Amidst a few tries, he was finally able to get the right one. He tapped his left ear with his left hand; listening in to the voice that poured through the mini speaker. "I just got new information." Key said, directing his words to Bell

"What?"

"A drone is coming in to help."

"A drone? Don't tell me… they are planning on dropping a missile bomb in the middle of the streets?" Bell gasped, shocked by the information he just heard. "How…? Isn't it too dangerous to do that???"

"Too late, the orders have already been sent out. Thirty seconds from now, it should be happening. We cannot lose the muscle cars. We must never lose them." Key replied, gritting his teeth as his hands tightened against the steering wheel. True to his words, the drone appeared thirty seconds later

"Brace for impact…" Key said, and almost immediately, the missile bomb dropped from above, landing on the car closest to theirs. The heat of the explosion could be felt as the car exploded, shrapnel from the explosion flying everywhere, some ricocheting off the exterior of the car. A bin near the affected car came flying towards them, creating an obstacle course for Key and Bell. Key took a sharp swerve to avoid the brunt of the explosion; the wheels screeching at the velocity of the turn. The car in front was on fire, but Key knew they couldn't stop. There was one more Vehicle they needed to be taken down. Taking a deep breath, Key cranked the speed higher, driving full force towards the explosion.

“We are fucking going to die.” Bell yelled, placing both of her hands against her ears. Key was different to the whole situation, his eyes was smoldering alongside the fire as he whipped through it. Key could only breath normally as they left the exploded car behind, their sights on the other vehicle.

“Is this even legal???” Bell yelled, holding on tighter to her seat as Key maneuvered the car through the streets. “They almost killed us too!” Bell complained.

“But at least one car is down… one more to go.” Key replied

Bell looked over at Key with bewildered eyes, her mouth slightly ajar at Key’s comment. “I swear you do not look like a crazy old man.” She whispered, shaking her head. “But surprisingly, you do.” Bell added, looking out the window to check the car debris they had left behind.

“Hey, look on the bright side... No civilians were hurt in the making of this movie.” Key chuckled.

“For some reason, I can see some similarities between you and mad max right now.”

“Mad Max?” Key asked, glancing over at Bell, and then back on the road.

“Why do you sound happy about that?” Bell asked, shaking her head at the absurdity of the whole situation.

"No... I just..." He chuckled, clearing his throat. "To be honest... you are starting to sound like a whining teen." he retorted.

"Hell no..." Bell glared back, raising her gun once more to shoot at the vehicle in front.

At the same time, the helicopter's sniper gun fires a shot, hitting a passenger in the white van in front. The radio intercom on the car's dashboard starts buzzing, drawing's Key attention, as he reaches over to take it.

"Detective Key, come in? Clear." The voice on the radio buzzed.

"Detective Key here... Clear."

"Confirm Johnny Jones is in the moving car?"

"Affirmative." Key replied, glancing over at Bell who was mouthing. *'I am not fucking okay.'*

"Can you block the exit of the white van in front?"

"Roger that."

"Our sniper on the Helicopter will try aiming for another shot. During the distraction, take over suspecting vehicle two."

"Roger that." Key replied, signaling the end of the call. The small speaker attached to Key's ear gave new information as the person on the other line began to countdown.

"You might want to grip tighter to your seat." Key said, causing Bell to eye him nervously. "Another missile bomb?" Bell asked testily.

"No…"

"Then? What do you" Before Bell could complete her sentence, Key cranked the engine further up, gaining the distance between both cars in less than a minute, before swerving to the right, blocking the white van exit. Because of the sharp turn, the muscle car had no other choice but to stop unless it would crash into Key's car. Few inches away, it screeched to a stop, groaning under the pressure of the force move.

The sudden move had also lurched Bell to the right, the seatbelt biting into her skin. "Arrgh! Fuck, warning next time! Warning!" She snapped.

Key bit his tongue, so he could stop himself from saying he had given a warning earlier.

"I am sure my ribs are bruised."

"What are they doing…?" Key whispered, highly suspicious about how calm the car in front of them us.

"Did you say something?"

"They are here." Key replied, tilting his head towards the car that was slowing down the van in front.

The police had also arrived, blocking the other exit of the white van. Both sides were ready to attack if need be; as they slightly stalked their target like a predator. The slight distraction was enough opportunity for the muscle van to act, and in a split second, the car crashed into Key's car

"Damn it! Those bastards! My Maverick!" Key cried, the force of the car slamming into them, pushing the car backwards with a loud screech. The driver of the white van throws something out of the window; the sound of metal hitting the ground with a soft clink, before the device slowly rolled under Key's car.

"Was that a bomb?" Bell panicked, her hand pressed down

"I don't know…" Key gritted, the loud scratch of the white van against the Maverick causing deep pains in his heart. And though the Maverick was in park, the white van was able to move it with enough force. Reaching for Bell's gun, Key raised it up to take aim, only to be stopped by Bell.

"Don't shoot. John is in that van, remember?"

"Fuck!" Key cussed, frustrated as he looked up to find the driver of the white van smirking through the side mirror, and he couldn't help but feel a foreboding sensation wash across his skin. He remembered the driver had thrown something earlier, and he could feel the hair along his arms rise.

"Think Key, Think… what did he throw. It definitely can't be a bomb." Key muttered his mind racing. Key discarded the idea of a bomb, because he knew their car would also be affected. Except it was a sub-lethal device…. Sub lethal? Something like a *stun grenade* with a timer? One of the wheels of Key's car pressed down on a clunk of metal; the bump fairly noticeable. And just like that, the grenade was triggered.

Upon detonation, the stun grenade emits an intensely loud bang of one seventy decibels, alongside a blinding flash of more than one million candela within five feet of initiation, causing immediate flash blindness. The sheer volume of the detonation also caused temporary deafness to every other person surrounding the muscle car. The loud sound of the stun grenade had disturbed the fluid in Key's ear, causing a loss of balance. The concussive blast had also sent little sparks flying, hitting the exterior of the cars around.

CHAPTER FORTY FOUR

"Fuck!" Key cussed out loud; his eyes gritty and his ears ringing. He was in pain, but he could only dawdle in it for too long. Groaning, he removed the seatbelt, at the same time reaching for the other gun tucked in his pants. "Are you okay?" He asked, turning to look at Bell. She had been real silent since the explosion, and he wondered if she was harmed. Her eyes were closed, and her head was tipped backward; her chin jutting out a little bit.

"Hey Bell? Are you alright?"

"No, I am not okay." She groaned, placing her left hand on her forehead. "Is the van still in front?" She asked, her eyes still closed.

"It is… Stay in the car." Key warned, checking the review mirror for any new addition. There was something about the whole scene that unsettles him. Perhaps it was because the van was still in place, or the fact that his years of experience knew something was up. Tapping the ear buds, Key wondered if there was any

"In Comms. Come in?"

The static sound of the radio filled his ears, and at that moment Key knew he was left with only two choices; to get out and confront the assailants, or wait for new orders. "Fuck it, I am going in." He cussed, removing the safety from the gun. The more time he wasted here, the more time the attackers had to cause bodily harm to John. He couldn't let that happen.

"I am coming with you." Bell responded, reaching for the door.

"Oh, no you don't. You stay here."

"Why?"

"I don't even know what is going on right now. It is better if I check it out first." He argued. A minute of silence ensued, before finally she gave in. "Fine." She sighed, reloading her gun. "But if anything happens, I am going out." She added.

Key bit the tip of his tongue, withholding his words. This was not the time to argue, neither was it a time to negotiate. "Hmm." He grunted, giving neither an answer nor a denial. "Wish me luck." He replied instead.

"Good luck." Bell answered, crossing her fingers.

Nodding his head, Key breathed in deeply, slowly opening the door. The air outside the car was a little much than he imagined. It was almost as if rain would fall, but he knew it was just the wind and nothing more. Focusing on the van in front of him, he crouched low; waddling each time he took a step. Key knew there was no way they weren't aware that he had gotten out of the car, and there was always the risk that they would drive off as soon as he came close enough; or they could shoot at him since he was at range that was disadvantageous. Walking with a strange tempo, Key switched lanes, crouching further as he hid behind their van. Counting from number one to ten in his mind, Key tried to see if there was anyone that would get out of the van.

And just as he predicted, whoever was seated next to the driver opened the door. He could feel his heart pounding as the footsteps drew nearer, his blood coursing with adrenaline and an invisible pressure that wrapped around his body. Raising the gun higher, Key slowly began to stand up, making sure his back was firmly pressed against the van, and also making sure he made little to no noise. In one smooth move, he changed the direction of the gun; using it as a different weapon as he punched the assailant hard on the face. Grunts filled with pain filled the tensed air; but Key wasn't waiting to find out anything. Grabbing the man by the neck, he pressed it further down, raising his left knee to connect with it. The force sent the man staggering backward, spit and blood flying everywhere as both knee and face meet. Key internally winces as he imagined someone's knee hitting his face. With one last move, he delivered the third and final strike; he hit the man by the side of the head, knocking him out. Everything looked too easy, but the truth was it was the result of all the numerous training and exercises he got.

The driver of the van noticed his partner's plight, and soon enough he did the unthinkable. He reversed sharply, trying to run down Key. Key was somewhat aghast. Right next to him was his comrade, and yet – he was trying to kill him. The driver of the van tried again once more, trying to run down Key; but they had all forgotten one fact. Key still had a backup soldier – Bell; who was trying hard to monitor what was going on from a distance. With enough precision to make a veteran soldier proud, she shot once more; the loud bang traveling across the trajectory of the wind. The bullet hit the back tire, deflating it at once, and throwing the driver off guard.

Running towards the other side, Key held on to the man's shirt through the window, holding tight.

"You fucking bastard!" Key gritted, pulling him out of the car, and away from the road. The man was unconscious, but breathing.

With enough force to force the door hinges off its hook, Key pulled at the handle. His blood ran cold as he realized what was going on. "Fuck! Fuck!" He swore harshly, his face contorting into a ball of anger. The sound of an extra pair of footsteps made him look up, only to find out that Bell had gotten out.

"What is wrong?" She asked, her gun raised in position; ready to shoot if there was any need.

Nodding his head, Key placed his hands akimbo. "We have been pranked" He replied.

Confused, Bell stepped forward, peering into the confines of the van. "Oh shit!" She cussed, a bitter taste feeling her mouth. She could understand why Key had gone so pale. There was no captive in the van. No John, no nothing. They had played them good.

But the question was, since when, where and how did they change vehicles without them knowing.

"Do you think they managed to escape when the grenade went off?" She asked, turning back to Key.

"No... No, I don't think so. The time frame is too short to escape with a hostage. And knowing John, he would try to delay them for as long as he could. But..." Key trailed off, running a hand through his hair. He couldn't complete his words, but his thoughts continued to flow. 'But John wasn't here.'

Looking up into the sky, the twinkling stars remained the same, unbothered and unconcerned with whatever was going on. Key frowned, realizing something else was wrong. "It's gone…" He muttered.

"What's gone?"

"The helicopter." He replied,

CHAPTER FORTY FIVE

Rewind time thirty minutes earlier.

There was something different about being embraced by the natural breeze. It slowly washed against your skin, teasing the hairs it cut across while cooling the temperature of your body. The moment was better than a thousand air cons. John was not someone that was sentimental. No he wasn't. But then again, witnessing a time period that could be end of the world left you feeling pretty much unsettled. Tipping his head backward, he stared up at the night sky, and the first thing he noticed was how still it was. It was like the world was at peace, and everything that was happening was only a dream. "How ironic." He muttered, his lips twisting into a wry smile.

But irony was far from over.

As soon as he muttered those words, headlights of a camouflaged van he had not seen switched on; and the light pierced through the darkness, blinding him temporarily. Squinting against the bright light, John wondered who it was. The play was just beginning, and the climax of the story was yet to be over. Automatically, the intensity of the headlights reduced, or perhaps it was because John's eyes had somewhat adjusted to the luminance. Whatever it was, John could see clearer. The van approached, and didn't stop moving until it was full inches away from where he stood. With enough accuracy to scare the soul out of his body, the driver of the van swerved right, startling John who moved backward. "What the fuck..." He cussed beneath his breathe, placing a hand against a chest as he tried to see who the terrible driver was. But that question remained unanswered as the door to the van swung open with enough force to make him flinch.

"Hey, what the fuck are you..." The words died on the tip of his tongue as he looked inside the van. John could feel the horror rip through his body, his eyes twitching, and his breath coming in short gasps. There were no words. Not a single letter uttered as he stared at the occupants of the car. There was no need to. He knew at that moment what he needed to do. Closing his eyes against the decision he had to make, he regretted not acting faster. But then again, what could he do? He scoffed. Opening his eyes, he nodded his head, and just as he took a step forward, he turned around one last time.

Life was Picasso. It painted coincidences like the master it was. And at that moment, it was painting a masterpiece. As soon as John turned around, his gaze caught the gaze of another. Although brief, it was enough. It was enough time to leave a vital piece of information.

As soon as John sat down, the door closed with a snap, and the Van wasted no time in making its getaway. The thoughts in John's head turned dark, and the hairs on his arm raised on its own accord; spreading Goosebumps across his skin. They had gotten him through his weakest link; the one person he never thought they would make a connection with.

They had gotten Rita.

In the confines of the van were two men, strapped and menacing. Their faces were covered, causing them to look like the average thieves instead of the murderous kidnappers they really were. But he wasn't fooled either ways. He knew everything was connected, and there was no such thing as a coincidence in situations like this. John tightened his fist so hard, his nails dug into his skin, drawing blood from both ends.

Rita was currently bound; her arms was tied up behind her back, her eyes were covered with a piece of clothing, her mouth was also covered with duct tape, but her were legs left untouched. He could see the fear in her eyes; the sides widened as she stared at him. As if to protest what her presence had done, she shook her head, her words coming out muffled behind the tape that covered her mouth. He knew it all. He knew what she was about to say. John had dreamed of meeting her; seeing her face and explaining the fact that disappearing was out of his control. He wanted to reach across, and wrap his hands around her trembling shoulders, and assure her everything would be alright. But he couldn't do any.

"You've gotten me. Please let her go." John began, breaking the momentarily silence.

"We got you either ways." One of the men snickered, causing his partner to snicker. John felt his insides burn, his chest aching as he realized they had no thoughts on letting her go.

"Please." He pleaded once more, hoping and silently praying that they let her go. Rita was innocent; an unfortunate encounter had gotten her caught in a crossfire she should have never witnessed.

"Either shut the fuck up, or get your mouth covered." One of the men growled, threatening to tie him down the exact same way they had done to Rita. John wondered how they had gotten to her. Did they

visit the school? Her home? How long had she been captured? Those questions ran around his head, but none of them were going to get answered.

Back to the present

John could feel the adrenaline swimming through his veins, causing his heart to pump harder. There was a reason for this. He just watched his last hope vanish into thin air. Due to some sick twist of fate, his kidnappers had a substitute van to take their place; misleading Key and sending them on a jolly ride that had no end. The men in the van cackled, amused by their little trick; but for John it was nothing but horror. Rita was more scared than she had been a second ago. She couldn't tell what was going, but she knew it couldn't be anything good. She berated herself for everything that happened. If only she had been careful… not a fool blinded by her own emotions; then they wouldn't be here.

"One second thoughts, perhaps you should take a nap."

"What-

But before John could ask what they were talking about, he felt a sting on his lap and he realized they were planning on drugging him. It all happened so fast, he didn't have enough time to think about it. One second he was staring ahead, his head filled with thoughts on how he would get away, and the next moment his thoughts became unclear, and his sight became blurry. Whatever they injected felt like a relaxant, and no matter how much he fought it, he had no control of his body. Slowly, black spots appeared before his eyes, before finally he gave in to oblivion.

CHAPTER FORTY SIX

Bill felt more nervous than excited. He was nervous to see his best friend after so many days. He was nervous to step out of the world and see what was left. But more than that, he was ready to face whatever was left. He was more than ready.

"Are you okay?" Jordan asked, forcing a smile.

"Yes… I am okay." Bill replied, shaking his head.

He wanted to comfort his dad, and assure him that everything would be alright. But nothing seemed to flow naturally. He wished there was something he could do or say. Nodding, Jordan took a step backward watching as they loaded Bill into the helicopter. He had a lot to say, but like his son – no words came out. It was a silence filled with a lot of thoughts and emotions that ran wild. As soon as the blades picked up, the wind increased, flapping at his shirt and sending his hair into disarray. He kept his expression the same as always, holding the smile in place until the helicopter was far up in the sky, a distance away. He continued to stay that way, until finally the helicopter was no longer in sight, and almost immediately, the smile on his face fell off.

He turned his back towards the sun, away from the spot Bill had just left. Away from it all.

It didn't take long for the helicopter to reach spot B; the hidden location John was supposedly at. Bill felt his palms become sweaty, a sign that he was nervous once more. As soon as the helicopter touched ground, his heart began to race, and his mouth felt dry. The building stood gloriously above, casting a shadow against the ground and sending high tech vibes to anyone who saw it.

But his moment of awe was short lived, as a familiar face ran up to him. "Bill" It was Marcus, his eyes widened as he stared at him. Bill felt a bit conscious sitting in a wheel chair, but he had no choice. "It's been a while." He greeted.

"Yea... Yes." Marcus replied, clearing his throat as he spoke. "Jordan?"

"He is fine." Bill nodded, rubbing his nape.

"Good to hear that." Marcus responded. Bill couldn't shake off the feeling that something was wrong. Marcus expression was more than an awkward reaction; it was filled with angst, and turned into anxiousness as he realized something more was amiss.

"Did something-

"John went missing." Marcus burst out, cutting short Bill's words.

"What? How? When?" Bill queried, his eyelids twitching as he stared at Marcus.

"I don't know...." He groaned, covering his face with both hands. "One minute we were trying to tap into his brain waves, and the next thing I know both John and Key are missing." He explained, pacing with each word he said. "Follow me. I need to show you something." Marcus added, pointing towards the opposite direction.

Silently, Bill follower behind, his face expressionless. His expression continued to stay that way even as he watched the CCTV cameras of how John had disappeared. "Will they find him?" He asked calmly, which was more disturbing than his normal reaction.

Marcus bit on his lower lip, not sure how to reply. But his silence spoke volumes. "I am sorry." He muttered.

"The brain waves… what was that about?"

"Oh…well, yes… it's uhm…" Marcus began, explaining what he meant. Bill continued to listen, drumming his finger at the tip of the chair. "Sixty five hundred… E31 and 13… Optimus?" He mumbled.

"Did you say something?" Marcus asked, pausing.

"Can I see the result of the brain waves?" Bill asked, dusting an invisible lint.

Marcus wasn't sure if became colder with everything that had happened to him, or perhaps he was trying to hide his feelings. Either ways, the new Bill was different. "Is something wrong?" Bill asked, raising a brow.

"No… just wondering if… never mind. This way…" Marcus said, leading Bill to a different office. In the other room was Victor, and another person he couldn't recognize.

"Bill!" Victor greeted enthusiastically, standing up as he sighted him. Although their greeting was brief, it was filled with enough warmth to relax Bill's facial expression a bit.

"This is the

Bill stared at the computer screen, a frown between his brows as he observed the pattern.

"Optimus." Bill replied, but this time clearer.

"Optimus? What's that?" Victor asked, unconsciously touching his chin.

"I will show you."

CHAPTER FORTY SEVEN

As soon as John opened his eyes, pain radiated in every single part of his body causing him to wince out in pain. It felt like a truck had run over him. "Ugh." He grunted out in pain, rolling slowly to his back. His chest was beating so fast; the movement noticeable from a distance away. He wanted nothing more but to sleep and forget all that was happening. But reality was too cruel. Memories of what had happened earlier came flooding in, and his instincts slowly penetrated through his foggy brain. "Where am I?" He muttered, sitting up. The room was empty, nothing in sight, and as bright as light. It was almost like he was left in a psych ward like the ones he had seen in the movies. "Hello?" He called out, his voice sounding hoarse. John felt so thirsty, and he knew a drop of water down his throat would feel like heaven at the moment.

Pulling his legs up to his chest, he scanned the room once more, only stopping when he realized a CCTV camera was mounted on one of the walls. Whoever it was, they were definitely watching him. Mustering enough strength into his legs, John slowly stood up, walking towards the camera.

"Hello?" He repeated once more, staring straight at it. "I know you can see me. Why am I here? What do you want?" Frustrated, John felt like punching something… anything. But his anger was temporarily dissipated as a voice replied back.

"Quite bold of you, don't you think?" The voice asked.

"Who are you?"

"Someone you should fear." The voice replied. John winced inwardly at his words, trying hard not to show his fear on his face. Even though it was only little, he wasn't going to allow the enemy's fear win over him.

"Where's Rita?"

"Who's Rita?"

The real panic began to set in and John wondered if they had done something to her. "The girl you brought in with me, where is she? You've already gotten me, let her go. Isn't that the rule of the game?"

"No one set the rule for the game I play." The voice answered resolutely, causing John to flinch. But this time, he couldn't hide his expression. Silence ensued in the room once more, and John was feeling more anxious than he was earlier.

He thought about all the things he could say or do… whatever he could do to save Rita, and get back to the facility.

"Hey… to whoever is listening… uhm… I have a deal for you."

A thunderous laughter filled the room, causing the muscles attached to his shoulder blades to stiffen. He felt ridiculed and humiliated for no reason, but John knew he had to press on, he couldn't afford to let his anger take over. "I have something to offer."

“What makes you think I need anything from you? Do you really think you have anything to offer?”

“Isn’t that the reason you brought me here? I am sure there was a reason why.” John pressed on. He wasn’t sure where all the confidence was coming from; perhaps false bravado, but it was all he had at that moment.

“Hmm…” The voice mumbled. “You’re so daring and rude at the same time, it’s so funny. For some reason, I want to give you a chance for you to tell me whatever you want to say. If your words do not amuse me, I will kill you right on the spot.” The voice said I a tone so light, it was a complete contrast to the danger and threat it carried.

John felt his spine stiffen, and he wondered if perhaps he should back out. But then again, Rita had gotten into this mess because of him. “We need to see face to face if I am to offer my words.”

“Alright, I will send someone to get you. Don’t even bother trying to run away. It is close to impossible here. And if by some bizarre interference, you’re able to escape- well… I guess you’ll be going alone, and leaving a corpse behind.” The voice warned referring to what would happen to Rita if he tried to escape. John bit down on his lower lip, nodding his head in agreement.

Three minutes later, the hidden door to the room opened up with a gentle push, and by the door way stood a man John had never seen. He looked more like a banker than a supposed gangster. And the contrast made him seem out of place. “Follow me.” The man said, adjusting his glasses, before turning around and heading the way he came from.

Hurriedly, John followed, scanning his environment as they passed by. Wherever they were, the facility was big; much bigger than the government owner facility he had been abducted from. He wondered who these people really were, and how they had managed to grow this big without the government interfering. Or was the government sponsoring them? The thought flashed through his mind and John knew that option wasn't entirely impossible.

This was the same set of people that failed to protect them even single time. At this point, it was almost as if they were in cahoots; seemed it was too suspicious that they managed to evade security and cause chaos all the time.

"We are here." The banker look alike spoke out, stopping just in front of a door. "Be respectful." He added. And John knew it was more of an advice than a threat. Bracing himself, he passed through the door, only to be met with several other days. The interior of this part of the building was completely different. Every high tech security measure John could think of was put in place. With each step he took, each of his movements were observed and recorded. There really was no way to escape; a maximum security level that could only be spotted on the white house.

Finally, after ten minutes of rigorous search, he had only a door left. The banker lookalike, knocked on the door, and almost immediately the voice he had heard earlier came from within.

"Come in." The voice said.

As soon as John came face to face with the man with the voice, all he did was observe. The man seemed familiar, but at the same time a stranger. He was sure he had never met him before, and John wondered if the familiarity stemmed from the continuous danger the man sent to them.

"You only have ten minutes." The man smirked, drumming his finger tips on the table. John wanted to ask him why he was so obsessed with killing them. Why he had gone too far to hurt them if they didn't know each other. But then again, he knew if he did- he would be killed without a second though.

"What do you think about me?" John asked.

"What?" The man queried, raising a brow up. "That was totally unexpected. But since you ask, I don't swing that way kid… and even if I did, I am not interested in young brats like you."

"No… I don't mean sexually." John corrected, feeling disgusted at the thought of being with a man as evil as the one before him. "I will work for you, and give you a prototype to the app."

"What makes you think I don't have the prototype already? Let me tell you something kid. My network is as good as the white house, maybe even better. Do you think I don't know all I need to know?" The man asked, raising his brow once more.

"My brain… uhm…my brain has a lot to offer. If you give me a chance. You won't regret it."

"Was that your offer?" The man asked, his fingertips inches midair away from the table.

John paused his lips, unable to answer. Nothing was going to work… he was going to die, and drag an innocent person with him.

Suddenly, a thought cane to his mind, and he realized that he might as well give it a try. "No, that's not all."

An unidentified car stopped by the side of the road, and moments later two teenage kids were seen coming out; a male and a female. As soon as the male closer the door, the car zoomed off, leaving them in the dust.

"Are you alright?" John asked, turning towards Rita. He couldn't read her expression, neither did she say a word. His nervousness came back in full force, and the guilt that ripped through his soul sent him staggering a bit. "I see you're alive." She finally said, breaking the silence. "You disappeared."

"I am sorry." He replied apologetically, unable to meet her eyes.

"That's it…? No explanation?" She queried.

This time, it was John's turn to be quiet. Where would he start? And how could he explain to her all that had happened. "I am sorry." He repeated.

"I don't want to fucking hear it. God…" She groaned, squatting down. "I thought you died." She mumbled, her words a bit muffled by her hands covering her face. "I am glad you're alive."

A myriad of expressions ran through John at that moment, and he could feel the emotion welling up in his throat. Also bending down, he placed a hand gently on her shoulder, causing her to look up. Their eyes stayed that way for a few minutes, until finally John broke the silence. “I will explain everything. Every single detail. Can you give me more time?” He asked, scanning her expression as he waited for answer.

Nodding, she wrapped her arms around his neck, hugging him close. Although she was confused, and had so many questions, she let it go for that moment. She was happy he was here.

Minutes after that, John was able to get a stranger to lend them their phone, and finally- he was able to call the facility for help. He was sent back, while Rita was sent to the hospital.

As soon as he went back, his eyes were drawn to the one that looked at him calmly, their gazes unwavering as they continued to stare at him. “Welcome back.”

John didn’t know how much he longed to hear those words until that moment. “Bill.” He muttered, his eyes getting misty. Without a second thought he ran forward, hugging him as soon as he was close enough. He stayed that way for a few seconds. And continued to remain that way for a few more minutes. He had missed his buddy.

EPILOGUE

Three months later.

The sun glowed fiercely on the teenage boys sitting silently in the park. To the third eye, they looked like any other kid. Nothing about their beings spelt out heroes, or saviors that prevented the end of the world.

"What are you think about?" Bill asked, frowning as a patch of sun hit him square in the face.

"The world." John replied, twiddling his thumbs.

"Yeah?"

John didn't tell Bill he was also thinking about his kidnapping. Mostly John remembered his face — the anger and hatred and evil as he realized his plan was not going to succeed. There is no sound to this memory. The traffic and background noise that must have existed are gone, edited out as superfluous. All that I have left is an image. An image that sporadically came to his fore mind when he was least expecting it. Perhaps he should go to rehab? He thought. After all, it was about time.

The memory in his head ran for about forty five seconds in, the man is getting frustrated. John could see it. The angst that filler his world every time he resisted. Everything is falling apart, and he knows it. The hardness on his face twists into anger and then rage, almost as if it is traveling along some sort of spectrum of hatred. This happens quickly, like he has to touch every base when all he wants to do is get to the end. Then he is there and the decision is made.

The man grabs for his gun. He needs both hands to chamber the bullet; his right holds the metal, the left works the lever. And now it is just him, waving his gun like he wants to destroy the world and press reset. For three seconds, an eternity when you're waiting on bullets, John felt stuck in the middle.

"John! John!" Bill yelled, his voice finally permeating John's thoughts. "Geez, I have been calling you since a minute ago." He grumbled.

"Sorry… I just." John sighed. "I have been thinking about the kidnapping a lot." He finally confessed, rubbing his nape. "Although nothing much happened… I just…" He laughed tersely, trying to cover up the awkwardness his words had brought.

Bill pursed his lips, worry etched on his features as he stared at John. "Should we both go to rehab?"

"I swear I was just thinking about it. It's a wonder how we haven't gone crazy yet… tipped over the edge or something."

"Maybe it's because we have too much holding us down… the love and care…"

"Maybe… yeah, you're right." John nodded. "But, I not going to lie. It sounds cringe." He added, biting his lower lip to stop from laughing.

"Shut up." Bill scowled. "What about Rita? Are you guys still okay?"

"Yeah… we… talked a lot. I told her everything." John confessed. "I am glad she believes me, because I felt crazy narrating a story that seemed too ridiculous to be true. But, it is what it is. She's also fine… still traumatized by the whole ordeal. But… she told me she is getting better. The amount of guilt I feel knowing I am the cause of that." John mumbled..

"Don't get me started." Bill groaned. "Every single person impacted by the App and the fact that Optimus was all the cause… makes me mad and guilty."

"Yeah… you know, sometimes… I wonder how many random strangers out there had to suffer because of the app. Be it losing a loved one, money, health, anything… I just… the guilt runs deep, not going to lie." John said, a sad smile on his lips.

"I agree." Bill sighed. "I feel just the same."

They sat in silence for a few more minutes, and each of the teenage boys thought about everything that had happened. Their minds was far matured and couldn't be compared to an average teenager. After all- they had seen it all.

"You remember that girl I used to work with?"

“Cat or Rose…” Bill replied, trying to remember her name.

“No, Bell.” John corrected.

“Oh… yeah. So what about her?”

“What do you think about her?” John asked, his gaze burning holes on Bill’s face.

“Uhm… she’s talented… I don’t know. What exactly are you asking?”

“If you are interested, I could connect-

“Hell no.” Bill scoffed, rolling his eyes. “I am good. I love my singleness”

“Come on man, you need to live a little. You’ve never had a girlfriend.”

“I am not dying because of that, and besides you’ve only had one too.” Bill retorted.

“At least I have one.”

“And having one doesn’t make you a relationship guru.” Bill snapped back, not giving John a chance to build his point.

“Ugh… I swear to God….” He groaned. “You know what? Do whatever you want.” He added, throwing his hand up in the air.

Bill chuckled at John's reaction, shaking his head at his aggravated expression. It was just a normal day…

A normal day?

His train of thoughts shocked him, and Bill realized something at that moment. No matter how tumultuous one's life was, the ending was truly unpredictable. A wistful smile cut across his lips as he thought about all that had happened. The scenes flashed through his mind like a camera roll; the good, the bad. His mind embraced them all. Although the bad still gave him nightmares sometimes, he knew everything that happened, happened for a reason. Smiling, he turned towards John who was also smiling at the moment. He knew he wouldn't trade anything for this moment.

Truly.

Those that disappeared, the unfinished story, the chaotic world... was that truly the end?

Made in the USA
Monee, IL
04 November 2026

67386188R10148